Natural Computing Series

Series Editors: G. Rozenberg (Managing)
Th. Bäck A.E. Eiben J.N. Kok H.P. Spaink
Leiden Center for Natural Computing

Advisory Board: S. Amari G. F
K.A. De Jong C.C.A.M. Gielen
L. Landweber T. Martinetz Z.
E. Oja G. Păun J. Reif H. Rub
H.-P. Schwefel D. Whitley E. V

T0073988
r

Springer

Berlin
Heidelberg
New York
Barcelona
Hong Kong
London
Milan
Paris
Tokyo

Alex A. Freitas

Data Mining
and Knowledge Discovery
with Evolutionary Algorithms

With 74 Figures and 10 Tables

 Springer

Author

Dr. Alex A. Freitas
Computing Laboratory, University of Kent
Canterbury CT2 7NF, UK
A.A.Freitas@ukc.ac.uk

Series Editors

G. Rozenberg (Managing Editor)
Th. Bäck, A.E. Eiben, J.N. Kok, H.P. Spaink

Leiden Center for Natural Computing, Leiden University
Niels Bohrweg 1, 2333 CA Leiden, The Netherlands
rozenber@cs.leidenuniv.nl

Library of Congress Cataloging-in-Publication Data
Freitas, Alex. A., 1964–
    Data mining and knowledge discovery with evolutionary algorithms/Alex A. Freitas.
      p.cm. – (Natural computing series)
    Includes bibliographical references and index.
    ISBN 3540433317 (alk. paper)
      1. Data mining. 2. Database searching. 3. Computer algorithms. I. Title. II. Series.
QA76.9.D343F722002
006.3–dc21   2002021728

ACM Computing Classification (1998): I., I.2, I.2.6

ISBN 978-3-642-07763-0 Springer-Verlag Berlin Heidelberg New York

Springer-Verlag Berlin Heidelberg New York,
a member of Springer Science+Business Media
http://www.springer.de

© Springer-Verlag Berlin Heidelberg 2002
Softcover reprint of the hardcover 1st edition 2002

Cover design: KünkelLopka, Heidelberg
Typesetting: Steingraeber, Heidelberg
Printed on acid-free paper   SPIN: 10986554   45/3111 GF- 5 4 3 2 1

This book is dedicated to all the people
who believe that learning is not only
one of the most necessary but also
one of the noblest human activities.

Preface

This book addresses the integration of two areas of computer science, namely data mining and evolutionary algorithms. Both these areas have become increasingly popular in the last few years, and their integration is currently an area of active research.

In essence, data mining consists of extracting valid, comprehensible, and interesting knowledge from data. Data mining is actually an interdisciplinary field, since there are many kinds of methods that can be used to extract knowledge from data. Arguably, data mining mainly uses methods from machine learning (a branch of artificial intelligence) and statistics (including statistical pattern recognition). Our discussion of data mining and evolutionary algorithms is primarily based on machine learning concepts and principles. In particular, in this book we emphasize the importance of discovering comprehensible, interesting knowledge, which the user can potentially use to make intelligent decisions.

In a nutshell, the motivation for applying evolutionary algorithms to data mining is that evolutionary algorithms are robust search methods which perform a global search in the space of candidate solutions (rules or another form of knowledge representation). In contrast, most rule induction methods perform a local, greedy search in the space of candidate rules. Intuitively, the global search of evolutionary algorithms can discover interesting rules and patterns that would be missed by the greedy search.

This book initially presents a comprehensive review of basic concepts from both data mining and evolutionary algorithms, and then it discusses significant advances in the integration of these two areas. It is a self-contained book, explaining both basic concepts and advanced topics in a clear and informal style.

My research on data mining with evolutionary algorithms has been done with the help of many people. I am grateful to my research students for their hard work. They had to carry out a time-consuming iterative process of studying a lot about both data mining and evolutionary algorithms, developing an evolutionary algorithm for data mining, performing several experiments to evaluate the algorithm, analyzing the results, refining the algorithm, etc. That was not an easy task. In particular, I would like to thank (in alphabetical order of first name) Celia C. Bojarczuk, Deborah R. Carvalho, Denise F. Tsunoda, Dieferson L.A. Araujo, Edgar Noda, Fabricio B. Voznika, Fernando E.B. Otero, Gisele L. Pappa, Marcus V. Fidelis, Monique M.S. Silva, Otavio Larsen, Rafael S. Parpinelli, Raul T. Santos, Roberto R.F. Mendes, and Wesley Romao.

I am also grateful to my colleague Dr. Heitor S. Lopes, with whom I have had many interesting discussions about evolutionary algorithms for data mining and have shared the supervision of several of the above-mentioned students.

In addition, I thank an anonymous reviewer for reviewing the manuscript of the book, my colleague Dr. L. Valeria R. Arruda for reviewing the first draft of chapter 11, and Prof. Dr. A.E. Eiben for reviewing chapter 12. I am also grateful to the team at Springer-Verlag for their assistance throughout the project.

Most of the research that led to the writing of this book has been carried out in two universities, both located in Curitiba, Brazil. More precisely, I was with CEFET-PR (Centro Federal de Educacao Tecnologica do Parana) in 1998, and since 1999 I have been with PUC-PR (Pontificia Universidade Catolica do Parana). I thank my colleagues from both institutions for creating a productive, friendly research environment.

Finally, during my research on data mining with evolutionary algorithms I have been partially financially supported by a grant from the Brazilian government's National Council of Scientific and Technological Development (CNPq), process number 300153/98-8.

Curitiba, Brazil
March 2002   Alex A. Freitas

Table of Contents

1 Introduction

"Computers have promised us a fountain
of wisdom but delivered a flood of data."
A frustrated MIS executive
[Frawley et al. 1991, p.1]

1.1 Data Mining and Knowledge Discovery

Nowadays there is a huge amount of data stored in real-world databases, and this amount continues to grow fast. As pointed out by [Piatetsky-Shapiro 1991], this creates both an opportunity and a need for (semi-)automatic methods that discover the knowledge "hidden" in such databases. If such knowledge discovery activity is successful, discovered knowledge can be used to improve the decision-making process of an organization.

For instance, data about previous sales might contain interesting knowledge about which kind of product each kind of customer tends to buy. This knowledge can lead to an increase in the sales of a company. As another example, data about a hospital's patients might contain interesting knowledge about which kind of patient is more likely to develop a given disease. This knowledge can lead to better diagnosis and treatment for future patients.

Data mining and knowledge discovery is the name often used to refer to a very interdisciplinary field, which consists of using methods of several research areas (arguably, mainly machine learning and statistics) to extract knowledge from real-world data sets.

There is a distinction between the terms data mining and knowledge discovery which seems to have been introduced (and popularized) by [Fayyad et al. 1996]. The term data mining refers to the core step of a broader process, called knowledge discovery in databases, or knowledge discovery for short. In addition to the data mining step, which actually extracts knowledge from data, the knowledge discovery process includes several preprocessing (or data preparation) and postprocessing (or knowledge refinement) steps. The goal of data preparation methods is to transform the data to facilitate the application of a given (or several) data mining algorithm(s), whereas the goal of knowledge refinement methods is to validate and refine discovered knowledge.

The knowledge discovery process is both iterative and interactive. It is iterative because the output of each step is often fedback to previous steps, as shown in Figure 1.1, and typically many iterations of this process are necessary to extract high-quality knowledge from data. It is interactive because the user, or more precisely an expert in the application domain, should be involved in this loop, to help in data preparation, discovered-knowledge validation and refinement, etc.

In this book we focus mainly on the data mining step of the knowledge discovery process, but the data preparation step will also be discussed in chapters 4 (about non-evolutionary data preparation methods) and 9 (about evolutionary data preparation methods).

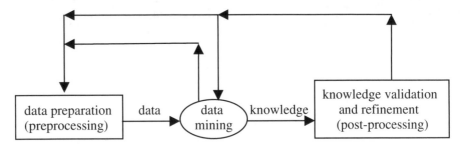

Figure 1.1: The iterative nature of the knowledge discovery process

1.1.1 Desirable Properties of Discovered Knowledge

Since the general goal of data mining is to extract knowledge from data, it is important to bear in mind some desirable properties of discovered knowledge. There is no consensus on what are the "ideal" properties of discovered knowledge, or even on how to define knowledge. One of the major problems is that to some extent knowledge, like beauty, is in the eye of the beholder.

Nonetheless, it is possible to mention at least three desirable properties of discovered knowledge. Namely, discovered knowledge should be accurate, comprehensible and interesting. The relative importance of each of these properties, which can be considered as quality criteria to evaluate discovered knowledge, depends strongly on several factors, such as the kind of data mining task (or problem) being solved, the application domain and the user.

In general, when the data mining task involves prediction, such as classification (section 2.1), it is important that the discovered knowledge have a high predictive accuracy, even though in many cases the comprehensibility and interestingness of the discovered knowledge tends to be more important than its predictive accuracy.

Knowledge comprehensibility is usually important for at least two related reasons. First, the knowledge discovery process usually assumes that the discovered knowledge will be used for supporting a decision to be made by a human user. If the discovered "knowledge" is a kind of black box which makes predictions without explaining or justifying them, the user may not trust it [Henery 1994]. Second, if the discovered knowledge is not comprehensible to the user, (s)he will not be able to validate it, hindering the interactive aspect of the knowledge discovery process, which includes knowledge validation and refinement.

The goal of discovering comprehensible knowledge can be facilitated by using a high-level knowledge representation, such as IF-THEN prediction rules (section 1.2.1). However, the use of such representation does not guarantee that the discovered knowledge is comprehensible. If the number of discovered rules and/or rule conditions is very large, the discovered knowledge can hardly be called comprehensible.

Finally, discovered knowledge should also be interesting, in the sense of being novel, surprising, and potentially useful for the user. This is probably the desired discovered-knowledge quality that most distinguishes data mining from the more "traditional" research areas on which it is based. In particular, the discovery of accurate knowledge has long been the goal of both machine learning and statistics. In general, the discovery of comprehensible, high-level knowledge has been less emphasized in statistics, but it has been considerably emphasized in some sub-areas of machine learning, which often use the above-mentioned high-level representation of prediction rules. In contrast, the discovery of interesting knowledge seems considerably more emphasized in the data mining literature than in the statistics and machine learning literature.

In any case, it should be noted that both rule comprehensibility and rule interestingness are in practice subjective concepts, difficult to effectively quantify, and there is a considerable controversy involving these concepts, particularly rule interestingness. The three rule-quality criteria of predictive accuracy, comprehensibility, and interestingness will be discussed in more detail in section 2.3.

1.2 Knowledge Representation

Among the several kinds of knowledge representation available in the literature, there are at least three that are often used in machine learning and data mining. [Langley 1996] calls them logical conjunctions, threshold concepts, and competitive concepts.

Logical conjunctions is the kind of knowledge representation used to express IF-THEN prediction rules, where the rule antecedent (IF part) consists of a conjunction of conditions and the rule consequent (THEN part) predicts a certain goal-attribute value for a data instance that satisfies the rule antecedent. In this book we focus on data mining algorithms that discover prediction rules expressed in this kind of knowledge representation. Hence, it will be discussed in more detail in subsection 1.2.1.

Unlike logical conjunctions, which are associated with an "all or none" matching between a concept description and a data instance, threshold concepts are associated with a partial matching between a concept description and a data instance. The basic idea is that a data instance satisfies the conditions of a concept description if it exceeds some threshold. This kind of knowledge representation is typically used in neural network algorithms, which are beyond the scope of this book. (We will only briefly mention the possibility of extracting rules from a neural network in section 1.3.)

Finally, the term competitive concepts also involves some form of partial matching, like threshold concepts. The main difference is that, rather than using a threshold to represent the necessary degree of matching, an algorithm computes the degree of matching for alternative concept descriptions and selects the best competitor. This kind of knowledge representation is typically used in instance-based learning (or nearest neighbor) algorithms. Although this paradigm of algorithm is not the main focus of this book, it is discussed in a few parts of this book, such as sections 3.3 and 9.2.

1.2.1 Prediction Rules

In this book we are particularly interested in the discovery of prediction rules. For our purposes a prediction rule is a knowledge representation in the form:

$$\text{IF } cond_1 \text{ AND ... } cond_i \text{ ...AND } cond_m \text{ THEN } pred.$$

This kind of rule consists of two parts. The rule antecedent (the IF part) contains a conjunction of m conditions about values of predictor attributes, whereas the rule consequent (the THEN part) contains a prediction about the value of a goal attribute. Presumably, predicting the value of the goal attribute is important for the user, i.e., an accurate prediction of that value will improve some important decision-making process. The semantics of this kind of prediction rule is as follows:

If all the conditions specified in the rule antecedent are satisfied by the predictor attributes of a given data instance – a record/tuple/row in the data set – then predict that the goal attribute of that instance will take on the value specified in the rule consequent. Some examples will be given below.

IF-THEN prediction rules are very popular in data mining because they usually represent discovered knowledge at a high level of abstraction, involving a few logical conditions. This knowledge representation is intuitively comprehensible for most users.

Each of the conditions in the rule antecedent can be in one of several different forms. The two most common ones are the following:

$$\text{(a) } Attr_i \text{ } Op \text{ } Val_{ij} \qquad \text{or} \qquad \text{(b) } Attr_i \text{ } Op \text{ } Attr_k$$

where $Attr_i$ and $Attr_k$ denote the i-th and k th attributes in the set of predictor attributes, Val_{ij} denotes the j-th value of the domain of $Attr_i$, and Op is a comparison operator – usually in $\{=, \neq, >, <, \geq, \leq\}$. The basic difference between these two forms of condition is that in the former we compare an attribute with a value belonging to its domain, whereas in the latter we compare two attributes. We will borrow some terminology from logic, and will refer to the former kind of condition as a propositional (or "0-th order") condition and to the latter kind as a first-order condition.

Two examples of propositional condition are: $(Sex = male)$ and $(Age > 21)$. We assume, of course, that the comparison operator makes sense for the domain of the attribute, so that conditions such as (*Sex* < *male*) are not valid. In general, the operator "=" is typically used when the attribute is categorical (nominal), whereas an operator such as "≤" or ">" is typically used when the attribute is continuous (real-valued). Propositional conditions are often called an attribute-value pair, even though the term is not very precise – since this kind of rule condition is specified by a triple attribute-operator-value, as mentioned above. In addition, propositional conditions can also refer to several values of an attribute, as, e.g., in the condition (*Marital_Status* IN {*Single, Divorced*}), which is equivalent to (*Marital_Status* = *Single* OR *Divorced*).

An example of first-order condition is: (*Income* > *Expenditure*). We assume, of course, that the attributes being compared have the same domain or at least have "compatible" domains. Actually, conditions in first-order logic can involve arbitrarily-complex predicates, rather than just comparison operators. These predicates can also have an arity (number of arguments) greater than two. For instance, a rule antecedent could have a condition such as *Supply(S,P,J)*, which would be true if a given supplier *S* supplied a given part *P* for a given project *J*.

First-order logic is much more expressive than propositional logic. We often need several propositional conditions to express the same concept as a single first-order condition. For instance, suppose that attributes *A* and *B* have the same domain, say, they can take on values in {v_1, v_2, v_3}. To express the first-order condition (*A* = *B*) in propositional form we would need a relatively long formula such as:

$$(A = v_1 \text{ AND } B = v_1) \text{ OR } (A = v_2 \text{ AND } B = v_2) \text{ OR } (A = v_3 \text{ AND } B = v_3).$$

The situation gets worse when we consider continuous (real-valued) attributes, such as *Income* and *Expenditure*. A simple first-order condition such as (*Income* > *Expenditure*) cannot be exactly represented by a finite number of propositional conditions.

However, the advantages of first-order logic representation do not come for free. The search space associated with first-order logic rules is usually much larger than the search space associated with propositional logic rules. In other words, in general the number of first-order-logic candidate solutions is much larger than the number of propositional-logic candidate solutions.

1.3 An Overview of Data Mining Paradigms

Since data mining is a very interdisciplinary field, there are several different paradigms of data mining algorithms, such as decision-tree building, rule induction, evolutionary algorithms, neural networks, instance-based learning (or nearest neighbor), bayesian learning, inductive logic programming and several kinds of statistical algorithms [Witten and Frank 2000; Dhar and Stein 1997; Mitchell 1997; Langley 1996; Michie et al. 1994].

It is difficult to make precise statements about the effectiveness of each of those paradigms in data mining applications, where the goal is to discover accurate, comprehensible, interesting knowledge from data. For one thing, each of these paradigms includes many different algorithms and their variations, with very different biases. The concept of bias will be discussed in detail in section 2.5. For now it is enough to say that each data mining algorithm has a bias, and any bias is suitable for some data sets and unsuitable for others. Hence, no data mining algorithm – and no paradigm – is universally best across all data sets.

In addition, even though most algorithms of a given paradigm may have a characteristic that is a disadvantage in the context of data mining, that characteristic can hardly be generalized to all algorithms of that paradigm. There are probably many exceptions. For instance, some people consider that neural networks (in general) are not suitable for data mining, because they work with a low-level knowledge representation, consisting of many numerical inter-connection weights. This kind of knowledge representation is not intuitive for the user, failing to satisfy the criterion of comprehensibility. However, there are nowadays many different methods to extract high-level, symbolic IF-THEN rules from neural networks [Taha and Ghosh 1999; Gupta et al. 1999], including evolutionary algorithms such as the one described in [Santos et al. 2000].

The focus of this book is on the paradigm of evolutionary algorithms, but we make no claim that this is the best data mining paradigm, for the above-mentioned reason that different algorithms have different biases, which are suitable for different data sets. Hence, anyone with an important data mining problem to be solved is well advised to consider the possibility of solving it with different data mining paradigms, trying to determine the best paradigm and the best particular algorithm for the specific target problem.

In the next two subsections we review the main aspects of two data mining paradigms, namely rule induction and evolutionary algorithms. The latter is the focus of this book, whereas the former is discussed here for the sake of comparison with the paradigm of evolutionary algorithms. This comparison is necessary because in this book we focus on evolutionary algorithms for discovering prediction rules, which is the same kind of representation of discovered knowledge in the rule induction paradigm. The rule induction paradigm will be reviewed in more detail in chapter 3, particularly section 3.2. That chapter will also contain a review of the decision-tree building paradigm (section 3.1) and a brief review of the nearest neighbor paradigm (section 3.3). The paradigm of evolutionary algorithms will be review in more detail in chapter 5.

Before we proceed, it is worthwhile to clarify an issue about terminology. In the next two subsections we discuss the paradigms of rule induction and evolutionary algorithms mainly in the context of rule discovery. In this narrow context, it might be said that rule induction and evolutionary algorithms are not two different paradigms, but rather two different ways of solving the problem of rule discovery. The rationale for this argument would be that both kinds of methods are based on the notion of searching a space of candidate rules. The difference between rule induction and evolutionary algorithms would be in the search strategy: the former uses a local, greedy search strategy, whereas the latter uses a kind of global, population-based search strategy inspired by natural selection.

This is one possible viewpoint, but in this book we prefer to consider rule induction and evolutionary algorithms as two distinct paradigms, for two main reasons.

First, we believe that the differences between rule induction-based search and evolutionary algorithm-based search are large enough to consider them as two different paradigms. These differences will be discussed in subsection 1.3.2.

Second, the notion of paradigm goes beyond the issue of specifying a search strategy to solve a rule discovery problem. Although our focus is on the use of evolutionary algorithms for rule discovery, in this book we also discuss how to use evolutionary algorithms to solve several other kinds of data mining problems, such as clustering (chapter 8), data preparation (involving attribute selection, attribute weighting, attribute construction, etc. – chapter 9), and tuning fuzzy membership functions (chapter 10). Actually, the notion of search paradigm goes beyond data mining. Evolutionary algorithms are a very general, flexible kind of search method, which are used in a number of other research areas.

Note that rule induction could also be considered a more general paradigm than its name suggests. The basic idea of rule induction is to perform a greedy local search in the space of candidate rules. If we generalize from candidate rules to candidate solutions, we have a general paradigm that can be used to solve other kinds of data mining problems, such as the above-mentioned data mining problems discussed in chapters 8, 9 and 10 of this book. Greedy local search is also a search paradigm which goes beyond data mining, being used in a number of other research areas. Of course, this also holds for several other "data mining paradigms", such as neural networks. We use the term "data mining paradigm" in this book to emphasize that we are interested in the paradigm's algorithms that are more suitable for data mining applications.

1.3.1 Rule Induction

In this paradigm the algorithm typically discovers knowledge expressed in the form of IF-THEN rules, as discussed in section 1.2.1. Rule induction algorithms can be naturally cast in terms of a state-space search, where a state is essentially a (part of a) candidate solution and operators (procedures) transform one state into another until a desired state is found. In the case of rule induction, a state corresponds to a candidate rule and operators usually are implemented by generalization/specialization operations that transform one candidate rule into another [Michalski 1983; Holsheimer and Siebes 1994]. After operators are applied to the current candidate rule(s), each new candidate rule produced by the operators is evaluated according to a heuristic function. In general only the best new candidate rule(s) is (are) kept, whereas the other new candidate rules are discarded. The algorithm keeps performing this iterative process until a satisfactory set of rules is found.

Figure 1.2 illustrates the concept of generalization/specialization operations. Figure 1.2(a) shows a candidate rule with two conditions, referring to the *Salary* and the *Age* of a customer, predicting that the customer will *Buy* a given product. This rule can be generalized by removing one of its conditions, as shown in the figure. This operation effectively generalizes the rule because the number of customers (data instances) satisfying the rule antecedent of the generalized rule is greater than or equal to the number of customers satisfying the antecedent of the original rule, since a rule antecedent is a conjunction of conditions.

Since specialization is the converse operation of generalization, one obvious way of specializing a rule is to insert one more condition into its antecedent. This operation effectively specializes a rule because the number of data instances satisfying the rule antecedent of the specialized rule is smaller than or equal to the number of data instances satisfying the antecedent of the original rule, since a rule antecedent is a conjunction of conditions. There are, of course, other ways of specializing a rule, which also have their generalization counterpart. For instance, Figure 1.2(b) shows that a rule can be specialized by making one of its conditions more strict, i.e., more difficult to be satisfied.

---

original rule: IF (*Salary* = *medium*) AND (*Age* ≤ 25) THEN (*Buy* = *yes*)
generalized rule: IF (*Salary* = *medium*) THEN (*Buy* = *yes*)

(a) Generalizing a rule by removing one of its conditions

original rule: IF (*Salary* = *medium*) AND (*Age* ≤ 25) THEN (*Buy* = *yes*)
specialized rule: IF (*Salary* = *medium*) AND (*Age* ≤ *18*) THEN (*Buy* = *yes*)

(b) Specializing a rule by making one of its conditions more demanding

---

Figure 1.2: Example of generalization/specialization in rule induction

As mentioned above, one way of implementing generalization/specialization operations consists of removing/inserting a condition from/into a conjunctive rule antecedent, respectively. Algorithms of the rule induction paradigm usually perform a local search for rules, constructing a rule by selecting one condition at a time to be inserted into or removed from a candidate rule. This makes this kind of algorithm relatively fast, at the expense of being quite sensitive to problems of attribute interaction, which may cause the algorithm to miss the best rule(s). In other words, a rule condition may seem irrelevant when considered separately, but it may become relevant when considered together with other conditions, in a conjunctive rule antecedent. This issue will be discussed in detail later in this book, particularly in subsection 3.2.1.

Concerning knowledge representation, most rule induction algorithms discover rules expressed in propositional logic (section 1.2.1), such as the rules shown in Figure 1.2. However, there are many rule induction algorithms that discover rules expressed in first-order logic [Lavrac and Dzeroski 1994]. In this book we will mainly focus on the simpler representation of propositional logic, but we will discuss the use of first-order logic representation in some parts of this book.

1.3.2 Evolutionary Algorithms

The paradigm of evolutionary algorithms consists of stochastic search algorithms that are based on abstractions of the processes of Darwinian evolution. The basic ideas of this paradigm are as follows. An evolutionary algorithm maintains a population of "individuals", each of them a candidate solution to a given problem. Each individual is evaluated by a fitness function, which measures the quality of its corresponding candidate solution. Individuals evolve towards better and better individuals via a selection procedure based on natural selection, i.e., survival and reproduction of the fittest, and operators based on genetics, e.g., crossover (recombination) and mutation operators. Hence, the better the quality of an individual, the higher the probability that its "genetic material" – parts of its candidate solution – will be passed on to later generations of individuals.

In essence, the crossover operator swaps genetic material between individuals, whereas the mutation operator changes the value of a "gene" (a small part of the genetic material of an individual) to a new random value. Both crossover and mutation are stochastic operators, often applied with pre-specified probabilities. Note that mutation can yield gene values that are not present in the current population, unlike crossover, which only swaps existing gene values between individuals. These operators will be discussed in detail in chapter 5.

Evolutionary algorithms are very flexible search methods. They can be used to solve many different kinds of problem, by suitably choosing an individual representation and a fitness function. In other words, one must decide what kind of candidate solution will be represented by an individual and which fitness function will be used to evaluate individuals. These choices are highly problem-dependent.

In this book we will focus on the use of evolutionary algorithms for discovering prediction rules, but other uses of evolutionary algorithms in data mining will also be discussed in some parts of this book. The main reason for our focusing on discovering prediction rules is the fact that this kind of knowledge representation tends to be intuitively comprehensible for the user, as mentioned earlier.

In the context of prediction-rule discovery, an individual corresponds to a candidate rule or rule set. The fitness function corresponds to some measure of rule or rule set quality. The selection procedure uses the fitness values of the individuals to select the best rules or rule sets of the current generation, whereas genetic operators transform a candidate rule or rule set into another rule or rule set.

We emphasize that, when evolutionary algorithms are used for rule-discovery purposes, they perform their search in a space of candidate solutions that is similar to the space searched by a rule induction algorithm. Actually, both kinds of algorithm will search the same rule space if they allow exactly the same kind of rule to be represented as a candidate solution. Hence, both kinds of algorithm have the advantage of discovering knowledge at a high level of abstraction, expressed as a set of IF-THEN prediction rules.

However, there are several important differences between the two kinds of algorithm concerning how they perform their search through the rule space. First, rule induction algorithms typically use deterministic operators. In addition, these operators usually perform a kind of local search in the rule space, in the sense that a single application of an operator modifies a small part of a candidate rule or rule set, say just inserting/removing a condition into/from a candidate rule. In contrast, evolutionary algorithms typically use stochastic operators. Some of these operators, such as crossover, usually perform a more global search in rule space, in the sense that a single application of an operator can modify a relatively large part of a candidate rule or rule set.

Second, rule induction algorithms typically construct and evaluate a candidate rule in an incremental fashion. In other words, many of the candidate rules evaluated by these algorithms are partial candidate rules. In contrast, evolutionary algorithms typically evaluate a complete candidate rule or rule set as a whole.

Third, unlike most rule induction algorithms, evolutionary algorithms work with a population of candidate rules or rule sets, rather than with a single rule or rule set at a time.

The above characteristics of evolutionary algorithms contribute for their performing a global search in the rule space, in contrast with the kind of local search performed by most rule induction algorithms [Freitas 2002a, 2002b; Dhar et al. 2000; Greene and Smith 1993]. As a result, intuitively evolutionary algorithms tend to cope better with attribute (or condition) interaction problems, as will be discussed in subsection 3.2.1.

On the other hand, a "pure" evolutionary algorithm approach suffers from a lack of task-specific knowledge, which is the kind of knowledge a rule induction algorithm typically has. For instance, a rule induction algorithm typically applies task-specific operators (e.g., generalization/specialization operations) to candidate rules in a systematic, directed manner, rather than applying general-purpose operators to candidate rules in a stochastic manner, as a pure evolutionary algorithm would do.

However, it is certainly possible to combine "the best of both worlds". Actually, one of the central themes of this book is to show how evolutionary algorithms can be adapted to use task-specific knowledge, in order to improve their performance in data mining tasks.

References

[Dhar and Stein 1997] V. Dhar and R. Stein. *Seven Methods for Transforming Corporate Data into Business Intelligence*. Prentice-Hall, 1997.

[Dhar et al. 2000] V. Dhar, D. Chou and F. Provost. Discovering interesting patterns for investment decision making with GLOWER – a genetic learner overlaid with entropy reduction. *Data Mining and Knowledge Discovery 4(4)*, 251–280, 2000.

[Fayyad et al. 1996] U.M. Fayyad, G. Piatetsky-Shapiro and P. Smyth. From data mining to knowledge discovery: an overview. In: U.M. Fayyad, G. Piatetsky-Shapiro, P. Smyth and R. Uthurusamy (Eds.) *Advances in Knowledge Discovery and Data Mining*, 1–34. AAAI/MIT Press, 1996.

[Frawley et al. 1991] W.J. Frawley, G. Piatetsky-Shapiro and C.J. Matheus. Knowledge discovery in databases: an overview. In: G. Piatetsky-Shapiro and W.J. Frawley (Eds.) *Knowledge Discovery in Databases*, 1–27. AAAI/MIT Press, 1991.

[Freitas 2002a] A.A. Freitas. Evolutionary algorithms. To appear in: J. Zytkow and W. Klosgen (Eds.) *Handbook of Data Mining and Knowledge Discovery*. Oxford University Press, 2002.

[Freitas 2002b] A.A. Freitas. A survey of evolutionary algorithms for data mining and knowledge discovery. To appear in: A. Ghosh and S. Tsutsui (Eds.) *Advances in Evolutionary Computing*. Springer, 2002.

[Greene and Smith 1993] D.P. Greene and S.F. Smith. Competition-based induction of decision models from examples. *Machine Learning 13*, 229–257. 1993.

[Gupta et al. 1999] A. Gupta, S. Park and S.M. Lam. Generalized analytic rule extraction for feedforward neural networks. *IEEE Transactions on Knowledge and Data Engineering, 11(6)*, 985–991, Nov./1999.

[Henery 1994] R.J. Henery. Classification. In: D. Michie, D.J. Spiegelhalter and C.C. Taylor. *Machine Learning, Neural and Statistical Classification*, 6–16. Ellis Horwood, 1994.

[Holsheimer and Siebes 1994] M. Holsheimer and A. Siebes. Data mining: the search for knowledge in databases. *Report CS-R9406* CWI, Amsterdam, Jan. 1994.

[Langley 1996] P. Langley. *Elements of Machine Learning*. Morgan Kaufmann, 1996.

[Lavrac and Dzeroski 1994] N. Lavrac and S. Dzeroski. *Inductive Logic Programming: Techniques and Applications*. Ellis Horwood, 1994.

[Michalski 1983] R. Michalski. A theory and methodology of inductive learning. *Artificial Intelligence 20*, 111–161, 1983.

[Michie et al. 1994] D. Michie, D.J. Spiegelhalter and C.C. Taylor. *Machine Learning, Neural and Statistical Classification*. Ellis Horwood, 1994.

[Mitchell 1997] T. Mitchell. *Machine Learning*. McGraw-Hill, 1997.

[Piatetsky-Shapiro 1991] G. Piatetsky-Shapiro. Knowledge discovery in real databases: a report on the IJCAI' 89 Workshop. *AI Magazine 11(5)*, 68–70, Jan. 1991.

[Santos et al. 2000] R. Santos, J.C. Nievola and A.A. Freitas. Extracting comprehensible rules from neural networks via genetic algorithms. *Proceedings of the 2000 IEEE Symposium on Combinations of Evolutionary Computation and Neural Networks (ECNN' 2000)*, 130–139. San Antonio, TX, USA. May 2000.

[Taha and Ghosh 1999] I.A. Taha and J. Ghosh. Symbolic interpretation of artificial neural networks. *IEEE Transactions on Knowledge and Data Engineering 11(3)*, 448–463, 1999.

[Witten and Frank 2000] I.H. Witten and E. Frank. *Data Mining: Practical Machine Learning Tools and Techniques with Java Implementations*. Morgan Kaufmann, 2000.

2 Data Mining Tasks and Concepts

> "...how should we generalize? How should we determine that we have
> enough instances of a generalization to warrant its acceptance?"
> [Holland 1986, pp. 231-232]

There are several data mining tasks. Each task can be considered as a kind of problem to be solved by a data mining algorithm. Therefore, each task has its own requirements, and the kind of knowledge discovered by solving one task is usually very different – and it is often used for very different purposes – from the kind of knowledge discovered by solving another task. Therefore, the first step in the development of a data mining algorithm is to define which data mining task the algorithm will address.

This chapter reviews the main data mining tasks addressed in this book and their corresponding main concepts. We start by reviewing in sections 2.1 and 2.2 the tasks of classification and dependence modeling, respectively. In both tasks discovered knowledge can be used for prediction. Indeed, chapters 6, 7 and 10 of this book focus on the discovery of prediction rules with evolutionary algorithms. Section 2.3 reviews some concepts about how to measure the quality of prediction rules.

Section 2.4 reviews the task of clustering. Evolutionary algorithms for clustering will be discussed in chapter 8. Finally, section 2.5 reviews the crucial concept of inductive bias of a data mining algorithm.

2.1 Classification

Classification is probably the most studied data mining task. We present below an overview of basic concepts and issues involved in this task. A more detailed discussion can be found in several good books about the subject, including [Hand 1997] and [Weiss and Kulikowski 1991].

In the classification task each data instance (or database record) belongs to a class, which is indicated by the value of a goal attribute. This attribute can take on a small number of discrete values, each of them corresponding to a class. Each instance consists of two parts, namely a set of predictor attribute values and a goal attribute value. The former are used to predict the value of the latter. Note that the predictor attributes should be relevant for predicting the class (goal attribute value) of a data instance. For example, if the goal attribute indicates whether or not a patient has or will develop a certain disease, the predictor attributes should contain medical information relevant for this prediction, and not irrelevant attributes such as the name of the patient.

Before we move on, a brief note about terminology seems appropriate. Since the classification task is studied in many different disciplines, there is a wide variety of terminology in use to describe the basic elements of this task. For example, a data instance can be called an example, an object, a case, a record, or a tuple. An attribute can be called a variable or a feature. In this book we will use mainly the terms data instance and attribute.

In the classification task the set of data instances being mined is randomly divided into two mutually exclusive and exhaustive (sub)sets, called the training set and the test set. The training set is made entirely available to the data mining algorithm, so that the algorithm has access to the values of both predictor attributes and the goal attribute for each data instance.

The aim of the data mining algorithm is to discover a relationship between the predictor attributes and the goal attribute using the training set. In order to discover this relationship, the algorithm has access to the values of both predictor attributes and the goal attribute for all instances of the training set. The discovered relationship is then used to predict the class (goal-attribute value) of all the data instances in the test set. Note that, from the viewpoint of the algorithm, the test set contains unknown-class data instances. Only after a prediction is made the algorithm is allowed to "see" the actual class of the just-classified data instance. If the class predicted by the algorithm was the same as the actual class of the instance, we count this as one correct prediction. If the class predicted by the algorithm was different from the actual class of the instance, we count this as one wrong prediction. One of the goals of the data mining algorithm is to maximize the classification accuracy rate in the test set, which is simply the number of correct predictions divided by the total number (correct + wrong) of predictions.

We emphasize that what is really important is to maximize the classification accuracy rate in the test set, rather than in the training set. Actually, the system could trivially "memorize" the training set, so achieving a classification accuracy rate of 100% in that set. This would involve no prediction. In contrast, when we use the discovered knowledge to classify unknown-class instances in the test set we are effectively doing a prediction, evaluating the discovered knowledge's generalization ability.

The above discussion was simplified, but for now it will do. A somewhat more detailed discussion on how to estimate classification accuracy will be presented in subsection 2.1.3. A discussion about other issues involved in evaluating the quality of discovered knowledge will be presented in section 2.3.

As mentioned in chapter 1, discovered knowledge can be expressed in many different ways. In this book we are mainly interested in the discovery of high-level, easy-to-interpret prediction rules, of the form:

IF (a_given_set_of_conditions_is_satisfied_by_a_data_instance)

THEN (predict_a_certain_class_for_that_data_instance).

To illustrate the above concepts, Table 2.1 shows an example of a very small training set with ten data instances (table rows), three predictor attributes – namely *Age*, *Gender* and *Salary* – and one goal attribute, namely *Buy*. In this example the aim of a classification algorithm is to discover rules that predict the value of *Buy* for data instances in the test set, by analyzing the training set shown in Table 2.1. Each instance corresponds to a customer of a hypothetical company. *Buy* can take on the values *yes* or *no*, indicating whether or not the corresponding customer will buy a certain product.

Table 2.1: Example of a training set given as input to a classification algorithm

Age	Gender	Salary	Buy (goal)
25	male	medium	yes
21	male	high	yes
23	female	medium	yes
34	female	high	yes
30	female	medium	no
21	male	low	no
20	male	low	no
18	female	low	no
34	female	medium	no
55	male	medium	no

One possible rule set that can be discovered from the training set of Table 2.1 is shown in Figure 2.1. The reader can check that the four rules shown in the figure are 100% consistent with the training set of Table 2.1, in the sense that, for all the training instances of Table 2.1 and for all the rules of Figure 2.1, whenever a given instance satisfies a given rule antecedent ("IF part"), the corresponding rule consequent ("THEN part") correctly predicts the class of that instance. However, 100% consistency with the training set does not guarantee that the discovered rules will have a high classification accuracy rate on the test set, consisting of instances *unseen* during training, as will be discussed later.

> IF (*Salary = low*) THEN (*Buy = no*)
> IF (*Salary = medium*) AND (*Age ≤ 25*) THEN (*Buy = yes*)
> IF (*Salary = medium*) AND (*Age > 25*) THEN (*Buy = no*)
> IF (*Salary = high*) THEN (*Buy = yes*)

Figure 2.1: Example of classification rules discovered from Table 2.1

The essence of a classification problem is illustrated, in a geometric form, in Figure 2.2. In this figure each data instance is represented by a point in the two-dimensional space formed by the predictor attributes A_1 and A_2. The coordinates of the point are given by the corresponding attribute values for the data instance. Each instance is labeled as "+" or "-", to indicate that it belongs to a "positive" or "negative" class, respectively. In this geometric interpretation, the goal of a classification algorithm is to find lines (or curves) that separate the data instances of one class from the data instances of the other class(es). Figure 2.2 shows the simple case where the data space is two-dimensional. More generally, in a data space with more dimensions the algorithm has to find hyperplanes that separate instances from different classes.

Figure 2.2(a) shows the original data set, without any line separating classes. Figure 2.2(b) shows one possible result of applying a classification algorithm to the data shown in Figure 2.2(a). In this case a single straight line has achieved an almost pure separation between the two classes. Figure 2.2(c) shows another possible result of applying a classification algorithm to the data shown in Figure 2.2(a). This time the algorithm used two straight lines to achieve a pure separation between the two classes. This does not necessarily mean that the partition of Figure 2.2(c) will have a better classification accuracy on the *test* set than the partition of Figure 2.2(b), as will be discussed in subsection 2.1.2.

Note that in the figure the lines used to separate the classes are straight lines orthogonal to the axes. Other kinds of line are used by some rule induction algorithms, but the kind of orthogonal lines shown in the figure is the most popular one, in the context of rule discovery. This is due to the simplicity of this representation, which creates data partitions that can be easily translated in terms of IF-THEN rules. For instance, the four data partitions shown in Figure 2.2(c) can be straightforwardly converted into the four rules shown in Figure 2.3, where t_1 is a threshold value belonging to the domain of A_1, and t_2, t_3 are threshold values belonging to the domain of A_2. In this example we are assuming that both A_1 and A_2 are numeric attributes.

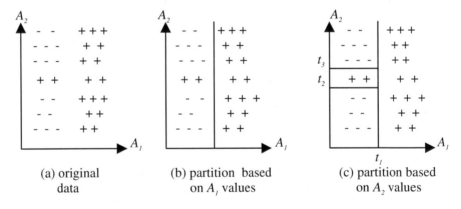

Figure 2.2: A geometric view of classification as a class-separation problem

IF $(A_1 > t_1)$ THEN $(class = $"+"$)$
IF $(A_1 \leq t_1)$ AND $(A_2 > t_3)$ THEN $(class = $"-"$)$
IF $(A_1 \leq t_1)$ AND $(t_2 \leq A_2 \leq t_3)$ THEN $(class = $"+"$)$
IF $(A_1 \leq t_1)$ AND $(A_2 < t_2)$ THEN $(class = $"-"$)$

Figure 2.3: Classification rules corresponding to data partition of Figure 2.2(c)

2.1.1 The Non-determinism
of the Classification Task

It is important to notice that classification is an *ill-defined, non-deterministic* task [Freitas 2000], in the sense that, using only the training data, one cannot be sure that a discovered classification rule will have a high predictive accuracy on the test set, which contains data instances *unseen* during training. In some sense, in classification we are essentially using data about "the past" (the training set) to induce rules about "the future" (the test set), i.e., rules that predict the value that a goal attribute will take on for data instances not observed yet. Clearly, predicting the future is a kind of non-deterministic problem.

Another way of understanding the non-determinism of classification is to recall that classification can be regarded as a form of induction, and that induction (unlike deduction) is not truth-preserving. To see why induction is ill-defined and non-deterministic, consider for instance the inductive task of predicting which is the next number in the following series: 1, 4, 9, 16, ?. (We suggest the reader actually spends a couple of minutes trying to predict the next number in the series, before moving on.)

The reader will probably have guessed 25, after inducing that the generator polynomium is n^2. However, the "correct" answer is 20, because the generator polynomium is: $(-5n^4 + 50n^3 - 151n^2 + 250n - 120) / 24$, borrowed from [Bramer 1996]. There are, of course, many other polynomia which could be the correct answer, since there is an infinite number of curves passing through a finite, small number of points. In other words, there is a virtually infinite number of hypotheses consistent with a training set, but the vast majority of them will make a wrong prediction on the test set. Clearly, we humans have a bias favoring the simplest hypothesis, but this is no guarantee that the simplest hypothesis will make the correct prediction – this point will be further discussed in section 2.5.2.

2.1.2 Overfitting and Underfitting

Let us revisit Figures 2.2(b) and 2.2(c), where all the data points are training data instances. Clearly, the partitioning scheme represented by Figure 2.2(c) has a 100% classification accuracy on the training set, while the partitioning scheme represented by Figure 2.2(b) does not, since in the latter the two positive-class ("+") data instances in the middle of the left partition would be misclassified as negative-class ("-") instances. The interesting question, however, is which of these two partitioning schemes will be more likely to lead to a higher classification accuracy on *unseen* data instances of the test set.

Unfortunately, there is no good answer to this question – based only on the training data, without having access to the test set. Consider the two positive-class data instances in the middle of the left partition in Figure 2.2(b). There are two cases to consider. On one hand, it is quite possible that these two data instances are in reality noisy data, produced by errors in collecting and/or inputting the data to the database system. If this is the case, Figure 2.2(b) represents a partitioning scheme that is more likely to have high classification accuracy on unseen data instances than the partitioning scheme of Figure 2.2(c). The reason is that the latter is mistakenly creating a small partition – corresponding to the rule IF $(A_1 \leq t_1)$ AND $(t_2 \leq A_2 \leq t_3)$ THEN (*class* = "+") – that covers only two noisy data instances, and so will likely lead to wrong predictions of unseen data instances. In this case one can say that this rule would be *overfitting* the data.

On the other hand, it is possible that those two data instances are in reality true exceptions in the data, representing a valid (though rare) relationship between attributes in the data. In this case the partitioning scheme of Figure 2.2(c) is more likely to have high classification accuracy on unseen data instances than the one of Figure 2.2(b). In this case one can say that the left partition of Figure 2.2(b), corresponding to the rule IF $(A_1 \leq t_1)$ THEN (*class* = "-"), would be *underfitting* the data.

In the above discussion noisy data was pointed out as a possible cause of overfitting. This is not the only possible cause of this phenomenon. Overfitting can also be caused by the use of a very small or sparse training set [Freitas and Lavington 1998, pp. 92-96] or by *oversearching* [Quinlan and Cameron-Jones 1995]. The basic idea of oversearching is that, if a rule induction algorithm considers a very large number of candidate rules, it will eventually find a rule that reflects a spurious (or "fluke") relationship, which is unlikely to be true in test data *unseen* during training. In more general terms, when you repeat an experiment a very large number of times, rare outcomes will eventually happen just due to chance. ([Orkin 2000, p. 9] whimsically calls this "the lottery principle". The fact that there are a very large number of players explains why there are lottery winners, despite the usually astronomical odds against winning lotteries offering big prizes.) A related issue is the problem of multiple comparisons in induction algorithms, which is discussed in detail in [Jensen and Cohen 2000].

The terms noisy data and spurious relationships should not be confused. The term noisy data usually refers to errors in the data. As mentioned above, these errors are often produced when collecting the data and/or inputting it to the database system. A simple example would be typing the value 52 for *Age* when the correct value is 25. The term spurious relationships usually refers to apparently-true data relationships that, in reality, are not statistically-significant enough to justify the formation of a prediction rule. Spurious rules may be discovered even when there is no noise in the data. Spurious rules may be discovered due to the use of a very small training set or oversearching, as mentioned above.

Whatever the cause of overfitting, the basic problem is that a rule overfitting the data will reflect idiosyncrasies of the current training set that are unlikely to occur in unseen test data, which tends to decrease the classification accuracy on the test set. We emphasize that although overfitting is much more discussed in the literature than underfitting, both are potential problems for a data mining algorithm.

As a final remark, we point out that although this section has discussed overfitting and underfitting in the context of classification, these phenomena can also occur in other data mining tasks involving prediction, including the task of dependence modeling – discussed in section 2.2.

2.1.3   Estimating Accuracy
via Training and Testing

In the beginning of section 2.1 we have emphasized the importance of using a random partition of the available data into a training and a test set in order to estimate the performance of the algorithm on an "unseen" data set, and we have defined the classification accuracy rate as the number of correctly classified instances in the test set divided by the total number of instances (either correctly- or wrongly-classified) in the test set. Actually, this measure of classification accuracy rate, computed on the test set, is an estimate of the true classification accuracy rate over the entire unknown distribution of data instances. In this section we elaborate on this issue.

We start by noting that, although we use a random partitioning of the original data into a training and test set to evaluate the performance of a classification algorithm, in practice the real data used for the actual evaluation of the algorithm will come later. In other words, in real-world applications the whole process of classification can be normally thought of as consisting of three phases, as follows.

First, the algorithm extracts some knowledge from the training set. Second, one measures the classification accuracy rate of that knowledge on the test set, which is independent from the training set – since these two data (sub)sets were produced by a random partition of the original data set. The measured classification accuracy rate is an estimate of the true classification accuracy rate of the algorithm over the entire unknown distribution of data instances. Theoretical or academic research usually stops at the end of this phase, but in real-world applications of data mining there is usually a third phase: in the future the algorithm will be used to classify truly new, unknown-class data instances – instances which were not available in the original data set and whose class is truly unknown for the user. In practice, it is the classification accuracy rate on this third data set, to be available only in the future, that will determine the extent to which the algorithm was successful.

Note that, when the time comes to classify this third, "future" data set, we can use the entire original data set – containing the previously-used training *and* test sets – as the new training set. This maximizes the amount of data to be used to discover the classification knowledge (or classifier) that will be employed in practice, when the future data set becomes available. At this point the future data set will take the role of the test set.

The above discussion of the first two phases involved a single random partition of the original data into a training set and a test set. This procedure, sometimes called hold-out, is simple and often used in large data sets. However, it has two disadvantages. First, one part of the data (usually about 2/3) is used only for training and the other part of the data (usually about 1/3) is used only for testing.

This reduces the amount of data effectively available for training and testing. Second, the classification accuracy is estimated based on a single random partition of the data, which is not very significant, from a statistical viewpoint. These problems are serious particularly when the original data set is small. A popular way of tackling these problems consists of using a cross-validation procedure, rather than a hold-out procedure.

In cross-validation, first of all the data is randomly divided into k mutually exclusive and exhaustive partitions (or folds), where k is a user-defined parameter. A value of $k = 10$ is often used, producing a 10-fold cross-validation procedure. Each fold should have approximately the same number of data instances. For instance, in 10-fold cross validation, each fold should have approximately 1/10 of the original data instances.

Then the classification algorithm is run k times. In the i-th run, $i = 1,...,k$, the i-th fold is used as the test set and the other $k - 1$ folds are temporarily merged and used as the training set for that run. Note that in cross-validation each data instance is used exactly once for testing and exactly $k - 1$ times for training. At the end, the estimate of classification accuracy rate computed by the cross-validation procedure is simply the arithmetic average of the k accuracy rates on the test set obtained in the k runs.

A particular case of cross-validation is the leave-one-out procedure, where k is set to the number of data instances in the original data set. Hence, in each run of the classification algorithm only one data instance is used as the test set and all the other instances are used as the training set. One advantage of leave-one-out is that it maximizes the amount of data used for training in each fold. One disadvantage is that it is very computationally expensive, so that it is practical only for small data sets.

Whatever the procedure used to partition the data into training and test sets, it is important to emphasize that this partition must be randomly made, so that the training and test sets of hold-out, or the folds of cross-validation, are statistically independent from each other. In practice, an additional restriction is often imposed in this random partitioning process. This restriction is called stratification, and it consists of imposing the restriction that in each data partition – the training and test sets of hold-out, or each fold of cross-validation – the proportion (or relative frequency) of data instances for each class is approximately the same as in the entire original data set. For instance, assume we have a data set with two classes, c_1 and c_2, where class c_1 occurs in 80% of the data instances and class c_2 occurs in the other 20% of data instances. Then, in each data partition of hold-out or cross-validation, classes c_1 and c_2 should occur with approximately the same respective proportions. In this case the procedure would be called stratified hold-out or stratified cross-validation.

2.1.3.1 A Note on Statistical Significance

The above discussion has emphasized the importance of estimating an algorithm's classification accuracy rate on an independent test set. However, recall that this is still just an estimate of the true accuracy rate over the entire unknown distribution of data instances. In particular, that estimate will depend on the particular makeup of data instances in the test set. In general two different test sets will produce two different estimates of the classification accuracy rate, i.e., the estimate has a variance.

So, an important question is: "how precise is the accuracy rate estimate?", i.e., how close it is to the true accuracy rate over the entire unknown distribution of data instances. In general, the larger the size of the test set, the greater the confidence that we have in the accuracy rate estimate (the smaller its variance), but in any case it is useful to quantify how precise the accuracy rate estimate is. This quantification can be done by calculating the confidence interval for a given level of statistical significance. The basic idea is as follows.

The first point to note is that when we measure the classification accuracy rate on an independent test set we are actually performing a random experiment, since the test set is a random sample of the original data. If we repeat this experiment many times, each time with a different random sample used as the independent test set, we would expect the accuracy rate to vary over the different test sets. The value of the accuracy rate on a particular test set is a random variable following a binomial distribution. As the number of experiments grows, the binomial distribution can be closely approximated by the normal distribution.

Let Acc_S be the classification accuracy rate of a classifier on an independent test set S, and let Acc_D be the true classification accuracy rate over the entire unknown distribution D. Note that Acc_S is an unbiased estimate of Acc_D, but the former is not a perfect estimate of the latter. We expect Acc_S to vary depending on the particular makeup of the test set S, as mentioned above. To quantify how precise the accuracy rate estimate Acc_S is, we compute its confidence interval for a given confidence level CL as follows (assuming a normal approximation to the binomial distribution):

$$Acc_S \pm z_{CL}.StdDev_S,$$

where z_{CL} is the value of a standard normal random variable associated with a desired confidence level CL and $StdDev_S$ is the standard deviation of the estimate Acc_S. Values of z_{CL} for different confidence levels (expressed in %) can be easily found in statistics textbooks. For instance, the values of z_{CL} for the confidence levels of 90%, 95%, 98% and 99%, assuming two-sided confidence intervals, are as follows:

Conference level CL:	90%	95%	98%	99%
value of z_{CL}:	1.64	1.96	2.33	2.58

The above formula for the confidence interval can be interpreted as follows: With approximately $CL\%$ probability, the true accuracy rate Acc_D lies in the interval:

$$Acc_S \pm z_{CL}.StdDev_S$$

Let us give a simple example of the use of a confidence interval. Suppose that the test set has $n = 500$ data instances, out of which 400 are correctly classified. Then $Acc_S = 0.8$ (80%) and $StdDev_S = 0.018$ (1.8%). This value of $StdDev_S$ was calculated for a hold-out procedure by the below formula [Weiss and Indurkhya 1998, p. 38; Weiss and Kulikowski 1991, p. 45], where n is the number of data instances in the test set S.

$$StdDev_S = \sqrt{(Acc_S \times (1 - Acc_S)) / n}.$$

Hence, the confidence interval for Acc_S at the, say, 95% level is: $0.8 \pm 1.96 \times 0.018 = 0.8 \pm 0.035$, which means that the true accuracy rate Acc_D lies in the interval 0.8 ± 0.035 with a probability of 95%.

A more detailed discussion about the estimation of classification accuracy rates can be found in some machine learning or data mining textbooks, such as [Mitchell 1997, chap. 5].

2.2 Dependence Modeling

In general, dependence modeling is a data mining task that involves the discovery of dependences among attributes. Note that the classification task, discussed in the previous section, also involves the discovery of attribute dependencies. There is, however, an important distinction between these two tasks. In classification there is a single goal attribute to be predicted, and in principle we are only interested in how the goal attribute depends on the other (predictor) attributes. Hence, dependencies between predictor attributes only, unrelated to the goal attribute, are not usually considered interesting by their own, at least in the sense that they are not useful for classifying new data instances, which is the essence of the classification task.

In contrast, in general the dependence modeling task involves a broader notion of dependence, and is usually associated with a much larger search space. We can think of different versions of this task. In one version the task is completely unrestricted with respect to the choice of attributes from which dependences are discovered and with respect to the "direction" of the dependence. In the context of prediction rule discovery (which is the focus of this book), this means that a discovered rule can contain any subset of attributes, and that the task is completely "symmetric" with respect to the attributes, in the sense that each attribute can occur in either the antecedent ("IF part") or the consequent ("THEN part") of the rule.

Note the difference from the classification task, which is very "asymmetric" with respect to the attributes, in the sense that the goal attribute and the predictor attributes can occur only in a rule consequent and rule antecedent, respectively.

In this book we are mainly interested in a somewhat more restricted version of the dependence modeling task, where, among all available attributes, just a few of them are considered valid goal attributes, i.e., attributes that can occur in a rule consequent [Noda et al. 1999]. The choice of goal attributes is performed by the user – in the same way that a single goal attribute is chosen by the user in a classification task. These goal attributes can occur either in a rule consequent or in a rule antecedent – but not in both parts of the same rule. This restriction is necessary, e.g., to avoid the generation of useless rules such as "IF $(A=1)$ AND $(B=2)$ THEN $(A=1)$", where the consequent is trivially true, since its prediction is a subset of the conditions of the rule antecedent. The other (non-goal) attributes can occur only in a rule antecedent. In general we require that the rule consequent have a single goal attribute-value pair, whereas the rule antecedent consists of a conjunction of conditions. More precisely, dependency-modeling rules have the form:

IF (a_given_set_of_conditions_is_satisfied_by_a_data_instance)

THEN (predict_the_value_of_a_goal_attribute_for_that_data_instance),

where the rule antecedent conditions can refer to both non-goal attributes and goal attributes that do not occur in the rule consequent. Note that different rules can have different goal attributes in their consequent; unlike the classification task, where all rules have the same attribute in their consequent.

2.2.1 Dependence Modeling vs Association-Rule Discovery

In its original, standard form, the task of association-rule discovery can be defined as follows [Agrawal et al. 1993]. Consider a data set where a data instance consists of a set of binary attributes called items. Each data instance represents a customer transaction, and each item of that transaction can take on the value *yes* or *no*, indicating whether or not the corresponding customer bought that item in that transaction.

An association rule is a relationship of the form X → Y, where X and Y are disjoint sets of items, i.e., X ∩ Y = ∅. Each association rule is usually evaluated by a support and a confidence measure. The support of an association rule is the ratio of the number of instances (transactions) having the value *yes* for all items in both the set X and the set Y divided by the total number of instances. The confidence of an association rule is the ratio of the number of instances having the value *yes* for all items in both the set X and the set Y divided by the number of instances having the value *yes* for all items in the set X. The association-rule discovery task consists of extracting from the data being mined all rules with support and confidence greater than or equal to user-specified thresholds (called minimum support and minimum confidence).

Prediction-rule discovery in the above-mentioned unrestricted, "symmetric" version of the dependence modeling task should not be confused with the association-rule discovery task. This latter is also "symmetric" with respect to the items (analogous to attributes) in the transactions (analogous to data instances).

However, association-rule discovery – at least in its standard form [Agrawal et al. 1993] – can be considered a relatively simple, deterministic task. In contrast, dependence modeling is a much more complex, ill-defined problem, which, like classification, can be considered non-deterministic. This issue and major differences between association-rule discovery and tasks involving prediction, such as classification and dependence modeling, are discussed in [Freitas 2000].

2.3 The Challenge of Measuring Prediction-Rule Quality

Measuring the quality of rules discovered by a data mining algorithm is a non-trivial problem, since this measurement can (and should) involve several criteria, some of them quite subjective. Ideally the rules discovered by a data mining algorithm should have three qualities, namely they should be accurate, comprehensible (simple), and interesting (novel, surprising, useful).

It is important to emphasize that a high mark in one or two of these criteria does not necessarily imply a high mark in the other(s). One very simple, reasonably popular example is appropriate here. Consider the following rule, that could be discovered from a medical database:

IF (*Pregnant? = yes*) THEN (*Gender = female*).

This rule has a very high predictive accuracy, and can be considered a simple, comprehensible rule. However, it is entirely uninteresting, since it contains a very obvious relationship.

The three above-mentioned rule-quality criteria are discussed in turn in the next three subsections. More precisely, subsection 2.3.1 discusses the problem of measuring predictive accuracy; whereas subsections 2.3.2 and 2.3.3 discuss the more difficult problems of measuring rule comprehensibility and rule interestingness, respectively.

2.3.1 Measuring Predictive Accuracy

In the context of prediction rules, it is very common practice to evaluate the quality of discovered rules with respect to their predictive accuracy. It is important to bear in mind, however, that this predictive accuracy must be measured on a separate test set, containing data instances that were *not* seen during training. This point was already made in section 2.1, when discussing the classification task, but it is worthwhile to emphasize some additional points.

First, the requirement of measuring predictive accuracy on a separate test set is not restricted to the classification task. It applies, in general, to any data mining task involving prediction, such as the dependence modeling task discussed in section 2.2.

Second, it should be noted that the challenge is really to achieve a high predictive accuracy in the *test* set, since achieving 100% "predictive" accuracy in the *training* set can be considered a trivial task. In the latter case, the algorithm just needs to "memorize" the training data. For instance, consider again the training set shown in Table 2.1. One can achieve 100% predictive accuracy in that training set by creating one prediction rule for each data instance, in such a way that: (a) the rule antecedent ("IF part") consists of the conjunction of all predictor-attribute values occurring in the instance; (b) the rule consequent ("THEN part") consists of the class of the instance. For instance, for the first data instance of Table 2.1 one would create the rule:

IF (Age = 25) AND ($Gender$ = $male$) AND ($Salary$ = $medium$) THEN (Buy = yes).

By creating 10 rules of this form, one for each data instance of Table 2.1, one would trivially achieve 100% of "predictive" accuracy in the *training* set. However, such a rule set will have no predictive accuracy in a *test* set containing instances different from the training instances. In other words, such a rule set involves no generalization at all. It is an extreme form of overfitting a rule set to the training data.

Third, as mentioned in section 2.1, a simple measure of predictive accuracy is the classification accuracy rate, which is the number of correctly-classified test instances divided by the total number (correctly-classified + wrongly-classified) of test instances. Although this measure is widely used, it has some disadvantages [Hand 1997]. In particular, it does not take into account the problem of unbalanced class distribution.

For instance, suppose that a given data set has two classes (goal-attribute values), where 99% of the data instances belong to class c_1 and 1% belong to class c_2. Such "extreme" class distributions are not so rare in practice as one might think at first glance. They are common in application domains such as fraud detection and diagnosis of rare diseases. The problem is that there is a trivial way to achieve a 99% predictive accuracy (in the test set) in this kind of data set. The algorithm just needs to classify all test instances with the majority class, i.e., the class with 99% of relative frequency. This does not mean that the algorithm would be doing a good job of predicting the class of test instances. It means that the above measure of predictive accuracy is too easy when the class distribution is very unbalanced, and a more challenging measure of predictive accuracy should be used in such cases.

More elaborate measures of predictive accuracy will be discussed in section 6.5 of this book, and are also discussed in more detail in [Hand 1997; Weiss and Kulikowski 1991].

Finally, it seems appropriate to mention here a couple of methodological mistakes that are sometimes made when measuring predictive accuracy on a separate test set. One methodological mistake consists of using the test set to "optimize" the parameters of a classification algorithm. In this case the algorithm is run many times, each time with a different set of parameter values, and the performance of the algorithm for each set of parameter values is measured on the test set. Then the set of parameters values with the best performance (on the test set) is chosen as the "optimized" setting of parameters. The flaw of this procedure is that it effectively uses the test set for training. Parameter setting must be done by using the training set only.

Another methodological mistake involves the issue of running a stochastic classification algorithm many times, with different values of random seed used for initializing the algorithm. This is typically done in the context of evolutionary algorithms (EAs). Multiple runs of an EA are of course desirable, to better validate the results produced by this kind of stochastic algorithm. However, in the context of data mining and prediction in general, a methodological mistake is made when only the accuracy rate (on the test set) of the best run of the EA, among all runs of the EA, is reported. For instance, suppose one runs an EA ten times, with ten different random seeds, and considers the best result over the ten runs (measured by accuracy rate on the test set) as the result of the EA. It is not fair to compare this result, produced by the best run of the EA, with the result of a single run of a non-stochastic rule induction algorithm.

After all, as discussed in subsection 2.1.2, the more a stochastic experiment (such as running an EA) is repeated, the larger the probability that rare outcomes (such as a high accuracy rate on a test set) will eventually happen just due to chance, rather than reflecting an ability of the algorithm in achieving a high accuracy rate. (This issue is discussed in detail by [Jensen and Cohen 2000].) When an EA is run many times, with different random seeds, one must consider the result of the EA as the average result on test set over all the EA runs.

Actually, if one reports the best result on the test set out of several independent runs, then one would be effectively making a mistake conceptually similar to using the test set for training. After all, in this case the EA would have the unfair benefit of classifying the test set several times in order to choose the best classification to be reported to the user.

2.3.2 Measuring Rule Comprehensibility

Knowledge comprehensibility is a kind of subjective concept – a rule can be little comprehensible for a user but very comprehensible for another user. However, to avoid difficult subjective issues, the data mining literature often uses an objective measure of rule comprehensibility: in general, the shorter (the fewer the number of conditions in) a rule, the more comprehensible it is. The same principle applies to rule sets. In general, the fewer the number of rules in a rule set, the more comprehensible it is. This is by far the concept of rule comprehensibility still most used in the literature, probably due to its simplicity and objectivity – one can easily count the number of rules and rule conditions and use this as a precise measure of rule comprehensibility. However, rule length is not the only factor influencing rule comprehensibility [Pazzani 2000; Freitas and Lavington 1998, pp. 13-14].

One of the main problems of relying on rule length alone to measure rule comprehensibility is that this criterion is purely syntactical, ignoring semantic and cognitive science issues. Intuitively, a good evaluation of rule comprehensibility should go beyond counting conditions and rules, and should also include more subjective human preferences. In particular, another factor influencing rule comprehensibility is the level of abstraction associated with the attributes occurring in the discovered rules. For instance, consider the following three ways of referring to the age of a person in a rule condition, in increasingly higher level of abstraction:

(a) testing whether the birth date of a person was after a given date, e.g., (*Birth Date after 06/06/1975*);
(b) comparing a person's age with a numerical threshold, e.g., (*Age < 26*);
(c) comparing a person's age with a pre-defined category, e.g., (*Age = young*).

It seems fair to say that representation (a) is the less comprehensible of the three above representations. The difference in comprehensibility between representations (b) and (c) is not so sharp, but most people would probably find representation of type (c) somewhat more comprehensible than the one of type (b). The difference is that in (c) the originally-continuous attribute was discretized, so that its numeric values were replaced but higher-level categorical values, such as *young, middle-aged, old*. Perhaps the difference is more noticeable in the case of an attribute which is "more continuous" than *Age*, say the attribute *Salary*. A rule condition such as (*Salary = high*) seems more comprehensible – more easily interpreted by a human user – than a condition such as (*Salary > $47,369.28*).

This is *not* to say that a higher-level, categorical representation is superior to a lower-level, numeric one. In many cases there is a trade-off between comprehensibility and predictive accuracy. When a categorical representation is produced by discretizing an originally-continuous attribute, it is well possible that this discretization leads to the loss of some relevant detail about the data; which can in turn decrease predictive accuracy, in comparison with the use of a finer-grained, continuous representation.

2.3.3 Measuring Rule Interestingness

Rule interestingness measures can be roughly divided into two groups: subjective (or user-driven) measures and objective (or data-driven) measures. Subjective measures are based mainly on taking into account previous knowledge or previous expectations of the user. When using this kind of rule interestingness measure, discovered knowledge is often considered interesting – in the sense of being novel and/or surprising for the user – when it contradicts the previous knowledge or expectation of the user.

In contrast, objective measures of rule interestingness try to estimate how interesting a rule will be for a user based mainly on the data being mined.

Intuitively, the main advantage and disadvantage of these two kinds of rule interestingness measure are as follows. On one hand, subjective measures have more direct and relevant information for selecting interesting rules for a given user. On the other hand, objective measures tend to be largely domain-independent, whereas subjective measures tend to be largely domain-dependent. In practice objective and subjective measures can be combined, rather than being used in a mutually-exclusive fashion.

Let us start with one representative measure of the class of subjective rule-interestingness measures.

[Liu et al. 1997] have proposed a subjective approach for selecting interesting rules, based on the notion of general impressions. In essence, the user specifies her/his general impressions about data relationships in the application domain.

These general impressions are specified in an IF-THEN, prediction-rule-like format. For instance, a given user might specify the following general impression: IF (*Salary* = *high*) THEN (*Credit* = *good*). Note that this is a *general* impression in the sense that it is quite vague (fuzzy), different from a reasonably-precise rule such as: IF (*Salary* > *$50,000*) THEN (*Credit* = *good*). The basic idea is that although the user is not supposed to know reasonably-precise rules, (s)he can have general impressions about the application domain that are a valuable clue for the system to determine what is interesting (novel, surprising) for the user.

Once the general impressions are specified, the discovered rules are compared with the general impressions. In a nutshell, there are essentially two kinds of interesting rules selected by the system. The first one consists of rules with unexpected consequent (THEN part). In this case the conditions in the discovered rule's antecedent (IF part) match a general impression's conditions, but the rule's consequent and the impression's consequent are different. Continuing the above example, suppose that the system discovered the rule: IF (*Salary* > *$50,000*) THEN (*Credit* = *bad*). This rule would be considered as interesting, since its consequent is unexpected (surprising) with respect to a general impression with the same antecedent. Another kind of interesting rule selected by the system consists of rules with unexpected conditions. In this case a rule's consequent matches a general impression's consequent, but the conditions in their antecedent are different.

We emphasize that the fact that a rule is interesting is no guarantee that it will be accurate and comprehensible. In the above example, there is no guarantee that the rule IF (*Salary* > *$50,000*) THEN (*Credit* = *bad*) is truly accurate. Presumably, the system would output this rule only if it estimated that this rule is not only interesting (novel, surprising) but also accurate. The estimate of accuracy is based on consistency with the data, whereas the estimate of interestingness is based on inconsistency with the user's general impressions. Therefore, when the above rule is shown to the user, presumably the user will carefully analyze it, and, simplifying the issue, there are two broad outcomes for the result of this analysis: (a) The rule is truly accurate, and the user's general impressions were wrong. In this case the system made a true discovery, contributing to improving the user's understanding of the application domain. (b) The rule is not accurate, and the system made a mistake. The cause of the mistake should be identified and corrected (e.g., the mistake might have been caused by some kind of noise in the data), so that the knowledge discovery process may be restarted.

In any case, the point is that a rule that is deemed by the system to be both accurate and inconsistency with the user's general impressions tends to be novel and surprising for the user. This kind of rule must be shown to the user, which should carefully analyze it to determine its true validity.

For a more comprehensive discussion about the general-impressions approach and related subjective measures of rule interestingness the reader is referred to [Liu et al. 1997, 2000; Liu and Hsu 1996]. In addition, a genetic algorithm designed for discovering interesting fuzzy rules based on the general-impressions approach is described in [Romao et al. 2002].

The general-impressions approach has the advantage of measuring rule interestingness in a direct way, giving the system access to the user's previous knowledge. However, it has two main disadvantages: (a) It requires that the user spends some time specifying general impressions. (b) It is application domain-dependent, i.e., the general impressions are valid only for the current application domain – and possibly only for the current user, since different users of a given application domain might have somewhat different general impressions about that domain.

These disadvantages raise the question: can one estimate that a rule is novel and/or surprising based only on objective, data-driven factors, without requiring that the user specifies her/his general impressions? If so, the system would have greater autonomy and generality, being to a large extent independent of the application domain.

The price to pay for this generality is that, intuitively, such an objective measure would be less effective in detecting rules that are really surprising for the user. After all, an objective approach does not have the benefit of being aware of the "previous knowledge" of the user, as specified by her/his general impressions. In any case, an objective rule interestingness measure that does a reasonably good job of (indirectly) estimating rule interestingness could certainly be useful. For instance, an objective rule-interestingness measure could be used as a kind of first filter to select potentially interesting rules, while other approaches (maybe based on visualization) would be used as a final filter to select truly interesting rules.

Several objective rule interestingness measures have been proposed in the literature. Here we first discuss the basic intuition about one approach for developing objective rule interestingness measures. Even if the system is not aware of the previous knowledge of the user, the system can make the educated guess that, probably, the user already knows strong, univariate (one-attribute-at-a-time) relationships between a predictor attribute and a goal attribute.

For instance, consider an application domain where the goal attribute is *Credit* and the set of predictor attributes includes the attribute *Salary* of a person. The user will probably know (or expect) that (*Salary* = *high*) is a condition normally associated with the prediction (*Credit* = *good*). Hence, a rule such as:

IF (*Salary* = *high*) AND ..... THEN (*Credit* = *good*)

tends not to be very surprising for the user – even though it might be overall a good rule, in the sense of being accurate and comprehensible. In contrast, a rule such as:

IF (*Salary* = *high*) AND ... THEN (*Credit* = *bad*)

will be surprising (and so potentially interesting) for the user, since its prediction is the opposite of what the user would expected, given the occurrence of the condition (*Salary* = *high*) in the rule antecedent. Note that although this rule seems inaccurate at first glance, it might be accurate due to an interaction of *Salary* and other attributes.

So far this example could be regarded as reinforcing the need for a subjective rule interestingness measure. After all, apparently the previous knowledge of the user has to be given to the system (possibly in the form of general impressions), in order to allow the system to infer that a rule such as

IF (*Salary* = *high*) AND ..... THEN (*Credit* = *bad*)

is surprising. However, it turns out that there is an objective, data-driven way for the system to make an educated guess about the user's previous expectations. In our example, one approach would be to analyze the class distribution associated with a rule condition. For instance, suppose that the condition (*Salary* = *high*) is satisfied by 100 data instances. If 90 out of these 100 instances have the class (*Credit* = *good*), the system can make the educated guess that the user will be surprised if (s)he is shown a rule such as:

IF (*Salary* = *high*) AND ...(other conditions)... THEN (*Credit* = *bad*).

In other words, the system can detect that (*Salary* = *high*) is correlated with (*Credit* = *good*), when the data is analyzed on an one-attribute-at-a-time basis. Therefore, if the system discovers a rule saying that (*Salary* = *high*) and other conditions are correlated with (*Credit* = *bad*), due to an interaction between *Salary* and other attributes, then this rule would be surprising for the user.

The above basic idea is used, in a much more elaborate form, in a method for summarizing and pruning discovered rules proposed by [Liu et al. 1999].

Two relatively-simple objective measures of rule surprisingness (or interestingness) were proposed by [Freitas 1998]. The basic idea of one of these measures is that a rule is considered surprising to the extent that it predicts a class different from the classes predicted by its minimum generalizations. Let a rule antecedent be a conjunction of m conditions, of the form $cond_1$ AND $cond_2$ AND ... $cond_m$. In essence, a rule has m minimum generalizations, one for each of its m conditions. The i-th minimum generalization of the rule, $i = 1,...,m$, can be obtained by removing the i-th condition from the rule antecedent. (Alternatively, a minimum generalization could be defined by relaxing a rule condition, rather than simply removing it.)

When a minimum generalization of a rule is generated, the system recomputes the class predicted by the generalized rule, which is the majority class of the data instances covered by the generalized rule. Let c be the class predicted by the original rule and let c_i be the class predicted by the i-th minimum generalization of the original rule. Then the system compares c with each c_i, $i=1,...,m$, and counts the number of times that c differs from c_i. The higher the value of this count, the higher the degree of surprisingness (interestingness) assigned to the original rule.

In other words, the system effectively considers that a rule has a large degree of surprisingness when attribute interactions make that rule cover a set of data instances whose majority class is different from the majority class of the sets of data instances covered by most of the minimum generalizations of that rule. One can also regard a rule with a large degree of surprisingness as an exception rule, since it covers fewer data instances and makes a prediction different from most of its minimum generalizations.

The second objective measure of rule surprisingness proposed by [Freitas 1998] is an information-theoretic measure. It starts by computing the well-known information gain of each predictor attribute – see, e.g., [Quinlan 1993, chap. 2]. In general, the higher the information gain of an attribute, the more relevant that attribute is to predict the class of a data instance, when attributes are considered individually, i.e., one at a time. However, it is likely that the user already knows what are the best predictors (individual attributes) for its application domain, and rules containing these attributes would tend to have a low degree of surprisingness for the user.

On the other hand, the user would tend to be more surprised if (s)he saw a rule containing attributes with low information gain. Although such attributes are irrelevant for classification when considered individually, one at a time, attribute interactions can render an individually-irrelevant attribute into a relevant one. If this is the case, a rule with low information-gain attributes not only could be an accurate rule, but also would tend to be a rule that is surprising for the user.

Therefore, all other things (such as the prediction accuracy) being equal, it can be argued that rules whose antecedent contain attributes with low information gain are more surprising (interesting, novel) than rules whose antecedent contain attributes with high information gain. This basic idea has been used in a GA designed specifically for discovering interesting prediction rules [Noda et al. 1999].

Several other objective, data-driven measures of rule surprisingness (or interestingness) and/or methods for discovering interesting rules are discussed in [Suzuki 1997; Suzuki and Kodratoff 1998; Suzuki and Zytkow 2000; Hilderman and Hamilton 1999, 2000; Piatetsky-Shapiro 1991; Freitas 1999].

It should be noted that several objective rule interestingness measures and/or methods for discovering interesting rules are based, implicitly or explicitly, in the notion of attribute interaction. In the context of the classification task, this notion can be intuitively defined as follows. Consider three attributes G, A_1 and A_2, where G is the goal (class) attribute to be predicted and A_1 and A_2 are predictor attributes. One can say that A_1 and A_2 interact when the direction or magnitude of the relationship between G and A_1 depends on the value of A_2. Actually, this can be called a two-way interaction. Higher-order attribute interactions can be defined in a similar way. A review of the role of attribute interaction in the discovery of interesting rules can be found in [Freitas 2001].

2.4 Clustering

In essence, the clustering task consists of partitioning the data being mined into several groups (or clusters) of data instances, in such a way that: (a) each cluster has instances that are very similar (or "near") to each other; and (b) the instances in each cluster are very different (or "far away") from the instances in the other clusters.

In other words, a clustering algorithm should maximize intra-cluster (or within-cluster) similarity and minimize inter-cluster (between-cluster) similarity. However, satisfying these two basic goals is not enough to obtain a good clustering solution, since these two goals can be trivially satisfied by simply assigning each data instance to a different singleton cluster. Therefore, it is also important to favor a relatively small number of clusters, increasing the number of data instances assigned to a cluster. A major challenge of the clustering task is to find a good trade-off between the above three goals.

In particular, in the context of data mining, which essentially involves searching for previously-unknown relationships in the data, it is desirable that the clustering algorithm be able to automatically determine the number of clusters and their shape, based on the data being mined [Ester et al. 1996].

It should be emphasized that there is an important distinction between two alternative goals of clustering, namely finding "natural" groups in the data and partitioning the data merely for convenience [Krzanowski and Marriot 1995]. The two goals are often pursued with similar clustering methods. However, in the latter case the number of clusters is usually chosen for convenience, while this choice is more difficult in the former case.

Formulating a precise, objective definition of a natural grouping of data is difficult, because this is a subjective concept. For instance, Figure 2.4 shows two distinct ways of clustering a two-dimensional data set with 13 data instances, labeled $A,...,M$. In Figure 2.4(a) the data is partitioned into three spherical clusters, whereas in Figure 2.4(b) the data is partitioned into two ellipsoid clusters. Both are valid clustering partitions, and it is difficult to say which of them is "better". It is clear, however, that both partitions are intuitively much better than many possible partitions for this data set. For instance, a partition that assign data instances A, B, C, D, L, M to one cluster and the other instances (E, F, G, H, I, J, K) to another cluster seems intuitively bad.

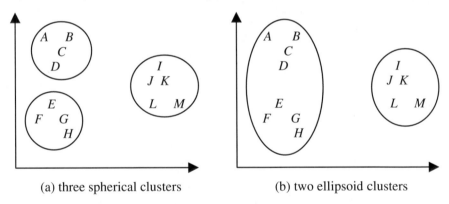

(a) three spherical clusters   (b) two ellipsoid clusters

Figure 2.4: Example of two distinct ways of clustering the same data set

Among the several types of clustering algorithms, we mention here two of the most popular ones: iterative partitioning (or partitional) and hierarchical clustering [Aldenderfer and Blashfield 1985; Backer 1995]. Each of these two types can be divided into two subtypes, as shown in Figure 2.5.

Hierarchical methods produce a hierarchy of clusters, whereas iterative-partitioning methods produce a "flat" clustering solution. Hierarchical methods can be subdivided into agglomerative and divisive methods. Agglomerative methods start assigning each data instance to one cluster, and then iteratively merge the two most similar (nearest) clusters until there is just one cluster, containing all instances of the data being mined (see subsection 2.4.1). Divisive methods work in the opposite way. They start assigning all data instances to one cluster. Then this cluster is iteratively divided into smaller and smaller clusters.

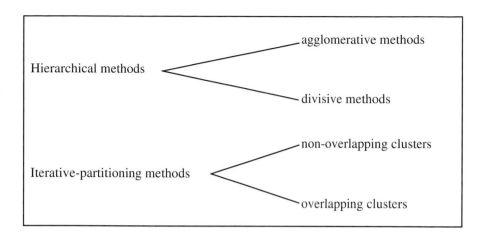

Figure 2.5: Two major types of clustering methods, each with two subtypes

Divisive clustering methods are in general much more computationally expensive than agglomerative methods. This is probably one of the reasons for the greater popularity of agglomerative methods.

Iterative-partitioning methods can be subdivided into methods that produce non-overlapping clusters and methods that produce overlapping clusters. In the latter a data instance can belong to two or more clusters at the same time, which is not allowed in the former.

Within each of the four subtypes of clustering algorithms shown in Figure 2.5 there are a large number of algorithms and their variations, whose review is beyond the scope of this chapter. We just present in the next two subsections a brief review of the basic ideas of hierarchical agglomerative clustering and one representative of the class of non-overlapping, iterative-partitioning clustering algorithms: the K-means algorithm. In particular, a basic understanding of the latter will be useful for a better understanding of some evolutionary algorithms for clustering to be discussed in chapter 8.

Finally, it is important to mention that some clustering algorithms are designed not only to discover clusters of similar data instances but also to discover some kind of pattern that specifies, in a generalized, abstract form, which data instances belong to each cluster. In this case the clustering task being solved is often called conceptual clustering. Examples of conceptual clustering algorithms can be found in [Fisher 1987] and [Talavera and Bejar 1998].

2.4.1 Hierarchical Agglomerative Clustering

The basic idea of an agglomerative clustering algorithm is quite simple, following a bottom-up approach. Let N be the number of data instances to be clustered. The algorithm starts with N clusters, each of them a singleton containing one of the original data instances. For every pair of clusters, the distance (or dissimilarity) between them is computed, according to a given distance measure. This produces an NxN distance matrix, whose cell (i,j) contains the value of the distance between clusters i and j. Then the algorithm merges the nearest pair of clusters, and a new $(N-1)$x$(N-1)$ distance matrix is formed. This process is iteratively performed until there is just one cluster, containing all the data instances of the original data set.

A very simple example of this basic idea is shown in Figure 2.6. Figure 2.6(a) shows a two-dimensional data space containing just four data instances – labeled A through D. Figure 2.6(b) shows the result of an agglomerative clustering algorithm applied to the data in Figure 2.6(a). This result is expressed in the form of a dendogram, presented with its "root" at the top.

It is easy to see how this dendogram was constructed. In the first step the algorithm merged the single-instance clusters A and B into the cluster (A,B), since this was the pair of nearest clusters in Figure 2.6(a). In the second step the algorithm merged the single-instance cluster C and the two-instance cluster (A,B) into the cluster (A,B,C), since the distance between C and (A,B) is smaller than the distance between (A,B) and D and smaller than the distance between C and D. Finally, in the third step the algorithm merged the single-instance cluster D to the three-instance cluster (A,B,C), creating a cluster that contains all four data instances.

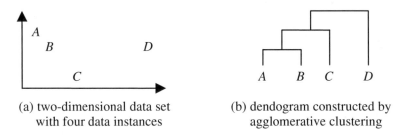

(a) two-dimensional data set   (b) dendogram constructed by
     with four data instances   agglomerative clustering

Figure 2.6: A very simple example of agglomerative clustering

2.4.2 The *K*-Means Algorithm

In this subsection we briefly review the K-means algorithm, a well-known iterative-partitioning clustering method. K-means partitions the data into K clusters, where K is a user-specified parameter. Each cluster is characterized by its centroid (or center). It starts with K centroids, and it iteratively performs the following two steps:

(1) assign each data instance to the cluster whose centroid is nearest to that instance;
(2) compute the new centroids of each cluster.

These two steps are iteratively performed until no data instance moves from other cluster to another.

The basic idea of the algorithm is that, since step (1) assigns each data instance to its nearest centroid, the summation of the distance between each instance and its nearest centroid is minimized. Note that the algorithm minimizes this total distance only for a fixed number (K) of clusters. This avoids the above-mentioned challenge of finding a good trade-off between the goal of maximizing discovered clusters' quality (maximizing intra-cluster similarity and minimizing inter-cluster similarity) and the goal of finding a relatively small number of clusters.

The K-means algorithm often produces a data partition that is a local optimum – rather than a global optimum – in the space of data partitions. This algorithm is also quite sensitive to the initial partition (the initial centroids), which is usually chosen at random. If it is important to find the global optimum, or at least a near-optimum, in practice the algorithm is run several times, with a different initial partition in each run.

2.4.3 The Challenge of Measuring Clustering Quality

We will see, in section 2.5, that no classification algorithm is universally best across all data sets. The same remark holds for clustering. The effectiveness of a clustering algorithm strongly depends on the shape and size of the "natural" clusters that are contained in the data set being mined [Aldenderfer and Blashfield 1984, pp. 59-60]. For instance, some methods favor the discovery of long-chained clusters – possibly described by (hyper)ellipses or (hyper)boxes – so that they are not the best method to use if the data contains spherical clusters.

Putting it in another way, the vast majority of clustering algorithms presupposes (implicitly or explicitly) some structure of the data, rather than inferring the structure of the data [Backer 1995, p. 124; Aldenderfer and Blashfield 1984, p. 16].

Unfortunately, one usually does not know the shape of the clusters hidden in the data a priori. Therefore, in practice a good strategy is to try several different clustering algorithms, with different biases, and pick up the best result – or some combination of the best results. It is also important to compare the results achieved by a clustering algorithm on a given data set with the results achieved by that algorithm on a *randomly-generated* data set having the same number of data instances and the same attribute domains as the original data set.

The above discussion can be summarized by the following remark of [Backer 1995, p. 149]:

"...there exists no single algorithm that may uncover all possible cluster structures. This problem can be solved by using the strengths of different methods, techniques and algorithms, aiming at carefully combining weighted pieces of evidence provided by them. Repeated experimentation, simulation with random data, and approximate reasoning are offered as the keystones of a combined approach."

2.5 Inductive Bias

A crucial concept in predictive data mining is the concept of inductive bias. An inductive bias can be defined as any criterion (explicit or implicit), other than strict consistency with the data, used to favor one hypothesis over another [Mitchell 1980].

In general, the term "hypothesis" denotes a candidate piece of knowledge being evaluated by the data mining algorithm. In the context of prediction-rule discovery, which is the focus of this book, a hypothesis would be a candidate prediction rule.

A very simple, pedagogical example of inductive bias can be found in [Schaffer 1993], as follows. Suppose that one tosses a counterfeit coin – where the probability of head is different from the probability of tail – 10 times and observes the outcomes (head or tail) of these tosses. Now consider the problem of predicting what will be the outcome of the next toss of the coin. An *unbiased* prediction strategy consists of predicting the outcome that has occurred more often in the previous 10 tosses. This prediction strategy takes into account only consistency with the observed ("training") data.

Consider now a strategy that predicts head if the outcome of at least 3 out of the previous 10 tosses was head. Clearly, this strategy has a *bias* favoring the prediction of head. The crucial question is: Is this bias good or bad? The answer is: it depends entirely on how the coin is counterfeit, which is precisely what we do not know and are trying to induce. If the counterfeit coin has a higher probability of head the above prediction strategy is good, but if the coin has a higher probability of tail the above prediction strategy is bad.

This example shows, in a very simple way, that the effectiveness of a bias is strongly dependent on the application domain. In other words, any bias will be suitable for some application domains (some data sets) and unsuitable for other application domains.

The need for bias becomes clear once one realizes that, given a set of data instances (training set), the number of hypothesis (e.g., candidate rules) implying these instances is potentially infinite [Michalski 1983]. Hence, any algorithm that discovers prediction rules must have an inductive bias. Without inductive bias there is no induction [Mitchell 1980; Mitchell 1997, p. 42], in the sense that the algorithm would be unable to prefer one hypothesis over many other hypotheses that are equally consistent with the data.

There are two main types of bias, as follows:

(a) Representation bias – The choice of a given representation language, in which candidate hypotheses (pieces of knowledge) are represented, introduces a representation bias into a data mining algorithm. The algorithm can only discover knowledge that is expressed in its representation language. For instance, if the algorithm's rule representation language allows rule conditions expressed in propositional logic but not in first-order logic, the algorithm will not be able to discover rule conditions that directly compare two attributes, such as *Income > Expenditure*, as discussed in subsection 1.2.1.

(b) Preference bias (or search bias) – This bias determines how the algorithm prefers one hypothesis (e.g., a candidate rule) over others. In the context of prediction-rule discovery, this bias is usually implemented by a function that evaluates the quality of a candidate rule and by the strategy used to generate new candidate rules from the current candidate rules.

2.5.1 The Domain-Dependent Effectiveness of a Bias

We emphasize that, as mentioned above, the effectiveness of any bias is strongly dependent on the application domain. Going somewhat further, this effectiveness is also strongly dependent on details of the particular data set being mined, among the numerous data sets that can be collected for any given application domain. In other words, claims such as "data mining algorithm A discovers prediction rules with greater predictive accuracy than algorithm B" should only be made for a given (or a few) application domains, or for a given (or few) data sets. In addition to being quite intuitive, taking into account the above discussion, the application-domain-dependent effectiveness of data mining algorithms has also been shown both theoretically and empirically. For a theoretical discussion the reader is referred to [Schaffer 1994] and [Rao et al. 1995]. (The paper by Rao et al. is a critical review of the paper by Schaffer. In particular, the work of [Schaffer 1994] is based on the unrealistic assumption that all possible assignments of classes to data instances of the test set are equally likely, regardless of the class assignments to data instances of the training set.)

It is worthwhile to mention here some empirical evidence of the application-domain-dependent effectiveness of data mining algorithms.

One comprehensive empirical comparison of different classification algorithms was carried out in the Statlog project [Michie et al. 1994; King et al. 1995]. This project compared the performance of about 20 classification algorithms in about 20 data sets. Out of several interesting conclusions of this project, one can mention, as an example, that decision tree-building algorithms tend to do well on credit data sets. One of the reasons seems to be that this kind of data set is usually produced by a human-decision maker who classifies the data considering the value of one predictor attribute at a time. Since decision trees classify the data in basically the same way (as will be explained in section 3.1), the bias of decision tree-building algorithms is naturally suitable for this kind of data.

Another interesting discussion about the application-domain-dependent effectiveness of data mining algorithms can be found in [Quinlan 1994]. This work distinguishes between parallel and sequential classification problems. In the former the vast majority of predictor attributes are relevant for predicting the class of a data instance. In contrast, in the latter the class of an instance depends only on a few predictor attributes, and the relevance of some predictor attributes depends on the values of other predictor attributes for the instance being classified. Then Quinlan conjectures, and shows some evidence for this conjecture, that a decision tree-building algorithm is unsuitable for parallel classification problems and that a well-known backpropagation neural network algorithm is unsuitable for sequential classification problems.

More recently, [Lim et al. 2000] have compared 33 classification algorithms of different paradigms on 32 data sets. More precisely, the 33 algorithms were 22 decision tree-building algorithms, 9 statistical algorithms and 2 neural network algorithms. With respect to classification error rate, overall the best algorithm was a statistical algorithm called POLYCLASS, which unfortunately does not produce comprehensible, IF-THEN prediction rules and requires a relatively long processing time. In any case, although POLYCLASS had the best (smaller) mean error rate across the 33 data sets, in only 15 data sets its error rate was within one standard error of the minimum for the data set [Lim et al. 2000, Table 4]. In addition, the error rate of POLYCLASS was *not* statistically significantly different from 20 other algorithms.

2.5.2   The Simplicity Bias of Occam's Razor

Occam's razor is a philosophical principle proposed by William of Occam in the 14th century. Translated into English, it essentially says that: "*Entities should not be multiplied without necessity*" or "Plurality should not be assumed without necessity" [Domingos 1998; Webb 1996].

Maybe never before in history a principle so simple and so influential in science has been so misinterpreted. More precisely, Occam's razor is often misinterpreted – not only in data mining, but also in science in general – as meaning: "*Among all hypotheses equally consistent with the evidence, the simplest one is the most likely to be true.*" Note carefully, however, that this interpretation does not follow from the original statement of the principle. Actually, while the original principle can be considered valid, this interpretation cannot.

Let us elaborate this point. Among all hypotheses equally consistent with the evidence, we should choose the simplest one, as recommended by the principle. However, the rationale for this decision is not that the simplest hypothesis is the one most likely to be true. After all, if all hypotheses are equally consistent with the evidence, all of them are equally likely to be true. Whether the simplest or most complex hypothesis will be the "true" hypothesis essentially depends on whether the real-world phenomenon being studied is in reality simple or complex. The rationale for choosing the simplest hypothesis, among all hypotheses equally consistent with the evidence, is that simplicity is desirable by itself – it is easier to work with simple hypotheses than with complex hypotheses. For instance, from a computational viewpoint, simpler hypotheses take less storage space, are faster to process, etc.

The same argument holds in the context of data mining, where the hypotheses are candidate pieces of knowledge (e.g., candidate prediction rules) being evaluated by the data mining algorithm, and consistency with the evidence means consistency with the data being mined. In this case, there are two versions of Occam's razor, which are called the sharp and the blunt by [Domingos 1998]:

The blunt: *"Given two models with the same training set error, the simpler one should be preferred because it is likely to have lower generalization error [on an unseen test set]."*

The sharp: *"Given two models with the same generalization error, the simpler one should be preferred because simplicity is desirable in itself."*

As implied by its name, the blunt version is not valid. Among all rule sets equally consistent with training data, all are equally likely to maximize predictive accuracy on an unseen test set. Whether the simplest or most complex rule set will really maximize predictive accuracy on an unseen test set essentially depends on whether the sought relationships in the data being mined are simple or complex. In other words, choosing a rule set because it is simpler than other rule sets that are equally consistent with the data is a form of inductive bias, i.e., a bias favoring simpler rule sets. As discussed in the previous section, any inductive bias is suitable for some application domains and unsuitable for others.

The sharp version, on the other hand, is indeed valid, because simplicity is desirable by itself. The simpler a rule set is, the easier a user can interpret it and assimilate it, the less computer memory it requires, etc.

For a more detailed discussion about this topic, we strongly recommend the reading of [Domingos 1998]. This is a very interesting paper, in which Domingos not only carefully deconstructs several arguments and well-known previous results that apparently support the blunt version of Occam's razor but also reviews several arguments and previous results that show that this version is not valid.

2.5.3 The Minimum Description Length (MDL) Principle

The Minimum Description Length (MDL) principle [Quinlan and Rivest 1989; Fayyad and Irani 1993] can be regarded as a heuristics used to discover knowledge that is both accurate and simple. Its basic ideas are as follows. Given a set of competing hypotheses (e.g., rule sets) and a data set, this principle recommends that one chooses, as the "best" hypothesis, the one that minimizes the sum of two terms, namely:

1) the length of the hypothesis; and
2) the length of the data given the hypothesis, i.e., the length of the data when encoded using the hypothesis as a predictor for the data.

The second term represents the length of the encoding of the data instances that are "exceptions" to the hypothesis. Both terms are measured in bits of information, which, one could argue, is a nice characteristic of the MDL principle. Apparently, it has the advantage that two incommensurable criteria for evaluating the quality of a hypothesis, its simplicity and its classification accuracy, are made "commensurable" by measuring them in terms of the number of bits needed to encode the hypothesis and the exceptions to the hypothesis.

However, one must be careful not to overrate this characteristic of the MDL principle. In particular, one should bear in mind the following points. First, the MDL principle introduces the problem of how to encode a hypothesis and its data exceptions into bits of information. Finding a "good" encoding scheme is often a difficult task – see, e.g., [Quinlan and Rivest 1989] – and the length of both the hypothesis and the data given the hypothesis is entirely dependent on this encoding. For instance, if there was reason to believe that complex hypotheses are more accurate than simple hypotheses, one could devise an encoding scheme where complex hypotheses had a shorter description in terms of bits of information. In this case the MDL principle would lead to the choice of accurate hypotheses that are simple in the devised encoding scheme, but complex in the original representation language of the hypothesis space. As [Domingos 1998, p. 38] insightfully points out:

"If they have higher priors [probabilities], more complex models can be assigned shorter codes, but this obviously does not imply any preference for simpler models in the original representation (e.g., if the model with the highest prior is a decision tree with a million nodes, it can be assigned a 1-bit code, without this implying any preference for small trees).
... Having assigned a prior probability to each model in the space under consideration, we can always recode all the models such that the most probable ones are represented by the shortest bit strings. However, this does not make them more predictive, and is unlikely to make them more comprehensible."

Now, suppose that we use a more "natural" encoding scheme, where the length of a hypothesis description in terms of bits of information is proportional to the length of the hypothesis description in the original representation language of the hypothesis space. In this case the use of the MDL principle would be favoring the discovery of a simple hypothesis in the original representation language of the hypothesis space. In other words, the MDL principle would be being used as an inductive bias favoring simpler hypotheses, and, as any other inductive bias, with respect to the maximization of classification accuracy the MDL principle would be appropriate in some application domains and inappropriate in others.

References

[Agrawal et al. 1993] R. Agrawal, T. Imielinski and A. Swami. Mining association rules between sets of items in large databases. In: P. Buneman, S. Jajodia (Eds.) *Proceedings of the 1993 International Conference on Management of Data (SIGMOD '93)*, 207–216. ACM Press, 1993.
[Aldenderfer and Blashfield 1984] M.S. Aldenderfer and R.K. Blashfield. *Cluster Analysis*. (Sage University Paper series on Quantitative Applications in the Social Sciences, No. 44.) Sage Publications, 1984.
[Backer 1995] E. Backer. *Computer-Assisted Reasoning in Cluster Analysis*. Prentice-Hall, 1995.

[Bramer 1996] M. Bramer. Induction of classification rules from examples: a critical review. *Proceedings of the Data Mining '96 Unicom Seminar,* 139–166. Unicom, London 1996.

[Domingos 1998] P. Domingos. Occam's two razors: the sharp and the blunt. In: R. Agrawal and P. Stolorz (Eds.) *Proceedings of the 4th International Conference on Knowledge Discovery and Data Mining (KDD '98),* 37–43. AAAI Press, 1998.

[Ester et al. 1996] M. Ester, H.-P. Kriegel, J. Sander and X. Xu. A density-based algorithm for discovering clusters in large spatial databases with noise. *Proceedings of the 2nd International Conference on Knowledge Discovery and Data Mining (KDD '96),* 226–231. AAAI Press, 1996.

[Fayyad and Irani 1993] U.M. Fayyad and K.B. Irani. Multi-interval discretization of continuous-valued attributes for classification learning. *Proceedings of the 13th International Joint Conference on Artificial Intelligence (IJCAI '93),* 1022–1027. 1993.

[Fisher 1987] D.H. Fisher. Knowledge acquisition via incremental conceptual clustering. *Machine Learning 2,* 139–172, 1987.

[Freitas 1998] A.A. Freitas. On objective measures of rule surprisingness. *Principles of Data Mining and Knowledge Discovery (Proceedings of the 2nd European Symp., PKDD '98) – Lecture Notes in Artificial Intelligence 1510,* 1–9. Springer, 1998.

[Freitas 1999] A.A. Freitas. On Rule Interestingness Measures. *Knowledge-Based Systems journal, 12(5–6),* 309–315, 1999.

[Freitas 2000] A.A. Freitas. Understanding the crucial differences between classification and discovery of association rules – a position paper. *ACM SIGKDD Explorations, 2(1),* 65–69. ACM, 2000.

[Freitas 2001] A.A. Freitas. Understanding the crucial role of attribute interaction in data mining. *Artificial Intelligence Review 16(3),* 177–199, 2001.

[Freitas and Lavington 1998] A.A. Freitas and S.H. Lavington. *Mining Very Large Databases with Parallel Processing.* Kluwer, 1998.

[Hand 1997] D.J. Hand. *Construction and Assessment of Classification Rules.* Wiley, 1997.

[Hilderman and Hamilton 1999] R.J. Hilderman and H.J. Hamilton. Heuristic measures of interestingness. *Principles of Data Mining and Knowledge Discovery (Proceedings of the 3rd European Conference, PKDD '99). Lecture Notes in Artificial Intelligence 1704,* 232–241. Springer, 1999.

[Hilderman and Hamilton 2000] R.J. Hilderman and H.J. Hamilton. Applying objective interestingness measures in data mining systems. *Principles of Data Mining and Knowledge Discovery (Proceedings of the 4th European Conference, PKDD '2000). Lecture Notes in Artificial Intelligence 1910,* 432–439. Springer, 2000.

[Holland 1986] J.H. Holland et al. *Induction: Process of Inference, Learning and Discovery.* MIT Press, 1986.

[Jensen and Cohen 2000] D.D. Jensen and P.R. Cohen. Multiple comparisons in induction algorithms. *Machine Learning 38,* 309–338, 2000.

[King et al. 1995] R.D. King, C. Feng and A. Sutherland. STATLOG: comparison of classification algorithms on large real-world problems. *Applied Artificial Intelligence*, 9(3), 289–333, 1995.

[Krzanowski and Marriot 1995] W.J. Krzanowski and F.H.C. Marriot. *Kendall's Library of Statistics 2: Multivariate Analysis – Part 2. Chapter 10 – Cluster Analysis*, 61–94. Arnold, London, 1995.

[Lim et al. 2000] T.-S. Lim, W.-Y. Loh and Y.-S. Shih. A comparison of prediction accuracy, complexity, and training time of thirty-three old and new classification algorithms. *Machine Learning 40*, 203–228, 2000.

[Liu and Hsu 1996] B. Liu and W. Hsu. Post-analysis of learned rules. *Proceedings of the 1996 National Conference of the American Assoc. for Artificial Intelligence (AAAI '96)*. AAAI Press, 1996.

[Liu et al. 1997] B. Liu, W. Hsu and S. Chen. Using general impressions to analyze discovered classification rules. *Proceedings of the 3rd International Conference on Knowledge Discovery and Data Mining (KDD '97)*, 31–36. AAAI Press, 1997.

[Liu et al. 1999] B. Liu, W. Hsu and Y. Ma. Pruning and summarizing the discovered associations. *Proceedings of the 5th ACM SIGKDD International Conference on Knowledge Discovery and Data Mining (KDD '99)*, 125–134. ACM Press, 1999.

[Liu et al. 2000] B. Liu, W. Hsu, S. Chen and Y. Ma. Analyzing the subjective interestingness of association rules. *IEEE Intelligent Systems 15(5)*, 47–55, 2000.

[Michalski 1983] R.W. Michalski. A theory and methodology of inductive learning. *Artificial Intelligence*, 20, 111–161, 1983.

[Michie et al. 1994] D. Michie, D.J. Spiegelhalter and C.C. Taylor. *Machine Learning, Neural and Statistical Classification*. Ellis Horwood, New York 1994.

[Mitchell 1980] T.M. Mitchell. The need for biases in learning generalizations. Rutgers Technical Report, 1980. Also published in: J.W. Shavlik and T.G. Dietterich (Eds.) *Readings in Machine Learning*, 184–191. Morgan Kaufmann, 1990.

[Mitchell 1997] T.M. Mitchell. *Machine Learning*. McGraw-Hill, 1997.

[Noda et al. 1999] E. Noda, A.A. Freitas and H.S. Lopes. Discovering interesting prediction rules with a genetic algorithm. *Proceedings of the Congress on Evolutionary Computation – 1999 (CEC '99)*, 1322–1329. Washington D.C., USA, 1999.

[Orkin 2000] M. Orkin. *What are the odds: chance in every day life*. W.H. Freeman, 2000.

[Pazzani 2000] M.J. Pazzani. Knowledge discovery from data? *IEEE Intelligent Systems,* 10–12, March/April 2000.

[Piatetsky-Shapiro 1991] G. Piatetsky-Shapiro. Discovery, analysis and presentation of strong rules. In: G. Piatetsky-Shapiro and W.J. Frawley (Eds.) *Knowledge Discovery in Databases*, 229–248. AAAI Press, 1991.

[Quinlan 1993] J.R. Quinlan. *C4.5: Programs for Machine Learning*. Morgan Kaufmann, 1993.

[Quinlan 1994] J.R. Quinlan. Comparing connectionist and symbolic learning methods. In: S. Hanson and J.R. Quinlan (Eds.) *Computational Learning The-*

ory and Natural Learning Systems: Constraints and Prospects, 445–456. MIT Press, 1994.

[Quinlan and Cameron-Jones 1995] J.R. Quinlan and R. Cameron-Jones. Oversearching and layered search in empirical learning. Proceedings of the 14th International Joint Conference on Artificial Intelligence (IJCAI '95), 1019–1024. Morgan Kaufmann, 1995.

[Quinlan and Rivest 1989] J.R. Quinlan. and R.L. Rivest. Inferring decision trees using the minimum description length principle. Information and Computation 80, 227–248, 1989.

[Rao et al. 1995] R.B. Rao, D. Gordon and W. Spears. For every generalization action, is there really an equal and opposite reaction? Analysis of the conservation law for generalization performance. Proceedings of the 12th International Conference Machine Learning (ML '95), 471–479. Morgan Kaufmann 1995.

[Romao et al. 2002] W. Romao, A.A. Freitas and R.C.S. Pacheco. A genetic algorithm for discovering interesting fuzzy prediction rules: applications to science and technology data. To appear in Proceedings of the 2002 Genetic and Evolutionary Computation Conference (GECCO '2002). New York, 2002.

[Schaffer 1993] C. Schaffer. Overfitting avoidance as bias. Machine Learning 10, 153–178, 1993.

[Schaffer 1994] C. Schaffer. A conservation law for generalization performance. Proceedings of the 11th International Conference Machine Learning, 259–265, 1994.

[Suzuki 1997] E. Suzuki. Autonomous discovery of reliable exception rules. Proceedings of the 3rd International Conference on Knowledge Discovery and Data Mining (KDD '97), 259–262. AAAI Press, 1997.

[Suzuki and Kodratoff 1998] E. Suzuki and Y. Kodratoff. Discovery of surprising exception rules based on intensity of implication. Principles of Data Mining and Knowledge Discovery (Proceedings of the 2nd European Symposium, PKDD '98). Lecture Notes in Artificial Intelligence 1510, 10–18. Springer, 1998.

[Suzuki and Zytkow 2000] E. Suzuki and J.M. Zytkow. Unified algorithm for undirected discovery of exception rules. Principles of Data Mining and Knowledge Discovery (Proceedings of the 4th European Conference, PKDD '2000). Lecture Notes in Artificial Intelligence 1910, 169–180. Springer, 2000.

[Talavera and Bejar 1998] L. Talavera and J. Bejar. Efficient construction of comprehensible hierarchical clusterings. Principles of Data Mining and Knowledge Discovery (Proceedings of the 2nd European Symposium, PKDD '98). Lecture Notes in Artificial Intelligence 1510, 93–101. Springer, 1998.

[Wcbb 1996] G.I. Webb. Further experimental evidence against the utility of Occam's razor. Journal of Artificial Intelligence Research 4, 397–417, 1996.

[Weiss and Kulikowski 1991] S.M. Weiss and C.A. Kulikowski. Computer Systems that Learn. Morgan Kaufmann, San Mateo, 1991.

[Weiss and Indurkhya 1998] S.W. Weiss and N. Indurkhya. Predictive Data Mining: a practical guide. Morgan Kaufmann, 1998.

3 Data Mining Paradigms

> "Learning algorithms in the rule-induction framework
> usually carry out a greedy search through the
> the space of decision trees or rule sets,..."
> [Langley 1996, p. 22]

As mentioned in the Introduction, since data mining is a very interdisciplinary field, there are many different paradigms of data mining algorithms, such as decision-tree building, rule induction, instance-based learning (or nearest neighbor), neural networks, statistical algorithms, evolutionary algorithms, etc. [Dhar and Stein 1997; Mitchell 1997; Langley 1996; Michie et al. 1994].

In this chapter we focus mainly on the decision-tree building and rule induction paradigms, which are discussed in sections 3.1 and 3.2, respectively. These two paradigms normally have an important advantage in the context of data mining, viz. they discover knowledge expressed in the form of a decision tree or a set of prediction rules. Both forms of knowledge representation are popular in the data mining literature, since they are intuitive for the user. As a result, the user can interpret and analyze the discovered rules, using them as a support for intelligent decision making, rather than blindly trusting on the output of a black box algorithm. Prediction rules are also the knowledge representation most discussed in several other chapters of this book. More precisely, the development of evolutionary algorithms for discovering prediction rules will be the central theme of chapters 6, 7 and 10, and a good understanding of rule discovery algorithms is also useful for a better understanding of chapter 9.

We also briefly discuss the instance-based learning (or nearest neighbor) paradigm in section 3.3, since several evolutionary algorithms for data preparation mentioned in chapter 9 involve the use of this kind of algorithm.

In general this chapter discusses the above-mentioned data mining paradigms in the context of the classification task (section 2.1), which seems the most addressed in the data mining literature and is also, overall, the most discussed in this book. However, the basic ideas of the data mining paradigms discussed in this chapter can also by used to develop data mining algorithms for other tasks. For example, the instance-based learning paradigm is often used to develop algorithms for the clustering task.

3.1 Decision-Tree Building Algorithms

In essence a decision tree is a knowledge-representation structure consisting of nodes and branches organized in the form of a tree such that: (a) every internal (non-leaf) node is labeled with the name of one of the predictor attributes; (b) the branches coming out from an internal node are labeled with values of the attribute in that node; (c) every leaf node is labeled with a class (a value of the goal attribute).

As a very simple example, let us assume that the decision tree shown in Figure 3.1 was built from the very small training set shown in Table 3.1. In this table each row corresponds to a customer. We will refer to a row as a data instance. The data set contains three predictor attributes, namely *Age*, *Gender* and *Salary*, and one goal attribute, namely *Buy*, whose value (to be predicted) indicates whether or not the corresponding customer will buy a certain product.

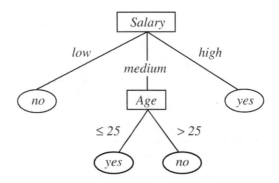

Figure 3.1: A decision tree built from the data in Table 3.1

Table 3.1: Data set used to build decision tree of Figure 3.1

Age	Gender	Salary	Buy (goal)
25	male	medium	yes
21	male	high	yes
23	female	medium	yes
34	female	high	yes
30	female	medium	no
21	male	low	no
20	male	low	no
18	female	low	no
34	female	medium	no
55	male	medium	no

We will discuss how to build a decision tree in the next subsection. For now let us see how a decision tree can be used to classify an unknown-class data instance. The basic idea is to push the instance down the tree, following the branches whose attribute values match the instance's attribute values, until the instance reaches a leaf node, whose class label is then assigned to the instance.

Suppose that the data instance to be classified is described by the tuple <*Age=23, Gender=male, Salary=medium, Goal=?*>, where "?" denotes the unknown value of the goal attribute. First the tree tests the *Salary* value in the instance. The answer is *medium*, so that the instance is pushed down through the corresponding branch and reaches the *Age* node. Then the tree tests the *Age* value in the instance. The answer is *23*, so that the instance is again pushed down through the corresponding branch. Now the instance reaches a leaf node, where it is classified as *yes*. Note that a decision tree does not need to consider all attributes of the instance, because some attributes (e.g., *Gender* in our example) may be irrelevant to a particular classification task.

3.1.1 Decision-Tree Building

A decision tree is usually built (or induced) by a top-down, "divide-and-conquer" algorithm [Quinlan 1993; Breiman et al. 1984]. Other methods for building decision trees are possible. For instance, genetic programming can be used to build a decision tree, as will be discussed in section 7.5.

Initialize T with the set of all training instances;
(1) **IF** all data instances in the current training set T satisfy a stopping criterion
    THEN
(2)   create a leaf node labeled with some class name and halt;
    ELSE
(3)   select an attribute A to be used as a partitioning attribute and
        choose a test, over the values of A, with mutually exclusive
        and collectively exhaustive outcomes $O_1,...,O_k$;
(4)   create a node labeled with the name of attribute A and create
        a branch, from the just-created node, for each test outcome;
(5)   partition T into subsets $T_1,...,T_k$, such that each T_i, $i=1,...,k$,
        contains all data instances in T with outcome O_i of the chosen test;
(6)   apply this algorithm recursively to each training subset T_i, $i=1,...,k$;
    ENDELSE
  ENDIF

Algorithm 3.1: A general top-down algorithm for building a decision tree

Here we briefly describe the general top-down approach for decision-tree building. Initially all data instances of the training set are assigned to the root node of the tree. Then the algorithm selects a partitioning attribute and partitions the set of instances in the root node according to the values of the selected attribute. The goal of this process is to separate the classes (values of the goal attribute), so that instances of distinct classes tend to be assigned to different partitions. This process is recursively applied to the instance subsets created by the partitions. A general algorithm to build a decision tree is shown in Algorithm 3.1 [Freitas and Lavington 1998], where T denotes the set of training instances in the current tree node. To facilitate our discussion, each step of the algorithm is identified by a sequential number between brackets on the left of the algorithm.

In step (1) the most obvious stopping criterion is that the recursive-partitioning process must be stopped when all the data instances in the current training subset T have the same class. In addition to this criterion, which can be considered as mandatory, several other stopping criteria can be used. One example is to stop the recursive-partitioning process when the number of data instances in T is below a small pre-specified threshold, on the grounds that there would be too few data instances to justify the creation of any new tree node.

In step (2) if all the data instances in the just-created leaf node have the same class the algorithm simply labels the node with that class. Otherwise, the algorithm usually labels the leaf node with the most frequent class occurring in that node.

In Step (3) the algorithm selects an attribute and a test over the values of that attribute in such a way that the outcomes of the selected test are useful for discriminating among data instances of different classes [Breiman et al. 1984, chap. 4; Quinlan 1993, chap. 2]. In other words, for each test outcome, the set of data instances that belong to the current tree node and satisfy that test outcome should be as "pure" as possible, in the sense of having many instances of one class and few or no instances of the other classes. In order to evaluate the purity of a data partition produced by a candidate attribute-test pair there are many different measures proposed in the literature [White and Liu 1994; Mingers 1989b]. Each of these measures has its own inductive bias (section 2.5), so that each measure is suitable for some data sets and unsuitable for others. In any case, when evaluating candidate attribute tests, the partitioning attribute can be either categorical or continuous. Let us discuss each of these cases in turn.

Partitioning based on a categorical attribute – If the partitioning attribute is categorical (nominal), a simple, common approach is to consider only candidate tests with one outcome for each of the distinct values of the partitioning attribute – see, e.g., the attribute *Salary* in Figure 3.1.

Alternatively, the algorithm can also consider tests where a single outcome can correspond to multiple values of the partitioning attribute [Quinlan 1993, chap. 7; Fayyad 1994], as long as the test outcomes are mutually exclusive and exhaustive. This approach is sometimes called subsetting in the literature. In the decision tree of Figure 3.1, for instance, the algorithm could consider the possibility of grouping the values *low* and *high* of the attribute *Salary* into a single branch. In this particular case this would not make sense, since these two attribute values lead to the prediction of opposite classes. However, this kind of grouping of attribute values can make sense when the attribute values being grouped lead to the prediction of the same class. In general this subsetting approach has the advantage of increased flexibility (see subsection 3.1.3) and the disadvantage of increasing processing time, by comparison with the simpler approach of associating a tree branch with a single attribute value.

Partitioning based on a continuous attribute – If the partitioning attribute is continuous (real-valued), the most common approach is to consider only candidate tests producing two outcomes, one for the partitioning-attribute values smaller than or equal a given cut point and the other one for the partitioning attribute values greater than that cut point – see, e.g., the attribute *Age* in Figure 3.1. This idea can be generalized to choose κ ($\kappa \geq 2$) cut points, so that $\kappa + 1$ outcomes are produced [Fayyad and Irani 1993]. This approach is sometimes called subranging in the literature. Similarly to the previously-discussed use of subsetting for categorical partitioning attributes, subranging increases flexibility at the expense of increasing processing time, since it increases the number of candidate partitions considered by the algorithm. [Quinlan 1996] discusses an elaborate way of handling continuous attributes in decision trees.

Steps (4), (5) and (6) of Algorithm 3.1 are self-explanatory. In addition to the above basic decision-tree-building algorithm, one must also consider the crucial issue of tree pruning, discussed in the next subsection.

3.1.2 Decision-Tree Pruning

Decision-tree pruning aims at producing simpler, shorter trees. One important motivation for tree pruning is that a shorter tree tends to be more comprehensible for the user. Another motivation is that an unpruned decision tree might contain some irrelevant nodes and/or be overfitting the training data, so that a pruned tree can, in some cases, obtain a better predictive accuracy on the test set. However, if tree pruning is too aggressive the tree will end up with the opposite problem, namely underfitting the data (subsection 2.1.2).

There are basically two approaches for decision tree pruning, viz. post-pruning and pre-pruning. In post-pruning the tree is expanded as much as possible by applying Algorithm 3.1 with a very demanding (difficult to be satisfied) stopping criterion in step (1) of the algorithm, and then the tree is pruned. In contrast, in pre-pruning partitioning can be stopped earlier during the recursive tree expansion process, if additional expansion of the tree does not seem promising. This corresponds to using a relaxed (easy to be satisfied) stopping criterion in step (1) of Algorithm 3.1.

Intuitively, post-pruning often produces trees with higher predictive accuracy than pre-pruning, since in the former the larger number of tree nodes provides more information for the pruning procedure. However, it should be noted that any pruning method is a form of bias [Schaffer 1993], so that its effectiveness (concerning its improvement in predictive accuracy) depends on the data being mined. On the other hand, pre-pruning makes the algorithm significantly more efficient, since it expands fewer tree nodes, which is appealing when applied to very large databases.

Out of the many tree-pruning methods proposed in the literature, we find it relevant to mention here a post-pruning method proposed by [Quinlan 1987a], whose variants are used in algorithms such as C4.5 [Quinlan 1993] and RULER [Fayyad et al. 1993]. The method first converts the tree to a set of IF-THEN classification rules. This conversion is straightforward. Each path of the tree going from the root to a leaf node is converted to a rule, as follows: (a) the internal nodes and their corresponding output branches are converted into conditions of the antecedent (the IF part) of the rule; and (b) the leaf node is converted into the consequent (the THEN part) of the rule. For instance, the decision tree shown in Figure 3.1 can be converted into the four rules shown in Figure 3.2.

> IF (*Salary* = *low*) THEN (*Buy* = *no*)
> IF (*Salary* = *medium*) AND (*Age* ≤ 25) THEN (*Buy* = *yes*)
> IF (*Salary* = *medium*) AND (*Age* > 25) THEN (*Buy* = *no*)
> IF (*Salary* = *high*) THEN (*Buy* = *yes*)

Figure 3.2: Rule set corresponding to the decision tree shown in Figure 3.1

Then, for each rule, the method iteratively removes the most irrelevant condition (the one whose removal most increases the rule's classification accuracy) as long as this removal process does not reduce the rule's classification accuracy.

This rule-pruning method has two main advantages. First, converting a decision tree into a set of rules often improves the comprehensibility of the discovered knowledge [Quinlan 1993, chap. 5], in the sense of reducing rule set size, which is often desirable in data mining. Second, this rule-pruning method considers removal of conditions regardless of the order in which conditions were generated, treating a rule as a set of conditions. This mitigates the problem of greedy attribute selection during decision-tree building, as will be discussed in subsection 3.1.3.

For a review of several decision-tree pruning methods the reader is referred to [Breslow and Aha 1997; Quinlan 1987b; Mingers 1989a; Esposito et al. 1993; Murthy 1998, pp. 358-360].

3.1.3 Pros and Cons
of Decision-Tree Algorithms

In this subsection we review the pros and cons of conventional decision-tree building algorithms. By "conventional" we mean in general algorithms which build a decision tree in a top-down fashion, by selecting one-attribute-at-a-time, as described in Algorithm 3.1.

From a data mining viewpoint, one advantage of this kind of algorithm is that it discovers knowledge that tends to be intuitively simple and comprehensible for the user, as long as the induced decision tree is not large. Large decision trees are difficult to interpret. Actually, to mitigate this problem some authors have proposed decision-tree pruning methods that simplify the induced tree even at the expense of reducing their predictive accuracy [Catlett 1991; Bohanec and Bratko 1994; Breslow and Aha 1997].

A decision tree also provides some information about the relative relevance of attributes for prediction purposes. The closer an attribute is to the root of the tree, the more relevant it tends to be for predicting the class of a data instance. This kind of information is not usually provided by "flat" (non-hierarchical) rule sets.

Another advantage of conventional decision-tree building algorithms is that they are fast [Lim et al. 2000], due to the use of a divide-and-conquer approach – which often works well in computer science. As the algorithm expands the tree the nodes contains fewer and fewer data instances. It should be noted, however, that this approach is a double-edged sword. On one hand, a smaller number of data instances implies a faster selection of attribute for the current node, since the attribute-selection procedure usually has a processing time proportional to the number of data instances in the current node. On the other hand, a smaller number of data instances means that there is less statistical support for the selection of an attribute. This is sometimes called the *fragmentation* problem in the literature [Friedman et al. 1996; Pagallo and Haussler 1990]. This problem can be somewhat mitigated by the use of the flexible subsetting approach, where a tree branch can be associated with multiple attribute values in step (3) of Algorithm 3.1, as explained in subsection 3.1.1.

Another problem associated with conventional decision-tree building algorithms is that they perform a greedy search for attributes to be put into the tree. By greedy search we mean this kind of algorithm builds a tree in a step-by-step fashion, adding one-attribute-at-a-time to the current partial tree, and at each step the best possible local choice is made. Note, however, that a sequence of best possible local decisions does not guarantee the best possible global decision. This limitation of greedy search is well-known, but it is still widely used in decision-tree building algorithms, due to its simplicity and computational efficiency. The problem of greedy search will be discussed in more detail in section 3.2.1.

A problem that seems to be less well-known is a problem associated with the structure of a decision tree. At each step of the tree expansion the algorithm must choose a partitioning attribute to label the current tree node and create branches coming out from that node. The branches must be associated with all possible values of the partitioning attribute. What if just one of those values is relevant for classification? For instance, perhaps the attribute-value pair (*Gender = female*) is relevant for classification, but the attribute-value pair (*Gender = male*) is irrelevant for classification. In this case, if *Gender* is chosen as a partitioning attribute the latter irrelevant attribute-value pair would be unduly included in the tree, to preserve the requirement that the branches coming out from a node must be associated with all possible values of the partitioning attribute. In cases where most attributes have just one relevant value, the inclusion of several irrelevant attribute values in a decision tree can considerably increase the size of the tree, making it unnecessarily complex [Cendrowska 1987].

The above two problems – associated with greedy search and decision-tree structure – can be somewhat mitigated by the use of a powerful rule-pruning method such as the one discussed in subsection 3.1.2. That rule-pruning method first converts the tree into a set of rules and then, for each rule, considers removal of conditions regardless of the order in which conditions were generated.

To see the importance of this point, suppose the attribute in the root of the tree is not very relevant for classification, i.e., assume that attribute was included in the tree due to the short-sightedness of the greedy decision-tree building algorithm. In order to remove the root node, most conventional post-pruning methods that stick to the tree representation would have to remove all nodes below the root node as well, which amounts to removing the whole tree. However, since the above-mentioned rule-pruning method converts the tree into a rule set and then prunes each rule separately, it can easily remove a condition that was in the root of the tree without removing any other condition of that rule and without affecting other rules [Fisher and Hapanyengwi 1993].

Despite its effectiveness, this rule-pruning method also has some disadvantages. For instance, it is computationally expensive, because the computation of statistics required to decide which condition should be removed from a rule often requires new scans of the training set. In addition, the entire approach can perhaps be considered somewhat cumbersome: it consists of first building a decision tree, then converting the tree into a rule set, and finally pruning those rules to achieve a more "flexible" rule set, which does not follow a decision-tree structure. This raises the following question: instead of using this approach, why not use an approach that discovers a more flexible rule set in the first place? Such an approach will be discussed in section 3.2.

The reader interested in critical reviews of decision tree algorithms is referred to [Bramer 1996; Fisher and Hapanyengwi 1993; Cendrowska 1987].

As a final remark before we move on to the next subsection, recently there has been active research on the development of variants of decision-tree building algorithms that are scalable to very large data sets. Some proposals and discussions on this topic can be found in [Gehrke et al. 2000; Srivastava et al. 1999; Lavington et al. 1999; Freitas and Lavington 1998].

3.1.4   The Inductive Bias of Decision-Tree Algorithms

It is important to understand the inductive bias of decision-tree building algorithms. That is, among the many decision trees consistent with a given training set, how do these algorithms choose one decision tree over the others? Answering this question is important to understand in which kind of data set this kind of data mining algorithm should be used. (Recall that any inductive bias is suitable for some application domains and unsuitable for others, as discussed in section 2.5.)

Many different decision-tree building algorithms have been proposed, with different inductive biases. However, at a high level of abstraction, one can say that most conventional decision-tree building algorithms have the following inductive bias: shorter decision trees are preferred over larger decision trees [Mitchell 1997, p. 63].

Therefore, roughly speaking, conventional decision-tree building algorithms have an inductive bias which is suitable to data sets where there is a relatively simple relationship between the predictor attributes and the goal attribute. In this case the relationship can be accurately represented by a relatively short tree. If the relationship in question is complex, it requires a large, complex tree for its accurate representation, so that the inductive bias of conventional decision-tree building algorithms tends to be unsuitable in this case [Schaffer 1993].

One example of a kind of data set where the simplicity bias of decision tree-building algorithms has produced good results was mentioned in subsection 2.5.1. The kind of data set in question was credit data sets used in the Statlog project. Apparently a major reason for the success of decision-tree-building algorithms in those data sets was that the simple approach of selecting one attribute at a time and favoring smaller trees was suitable for those data sets, which seem to have been produced by a human-decision maker who also classified the data considering one attribute at a time.

3.1.5 Linear (Oblique) Decision Trees

A linear (or oblique) decision tree is a decision tree where the tests on attribute values performed at each tree node are linear combinations of some of the (numeric) attributes. Therefore, the nodes represent oblique cuts in data space, in contrast with the axis-parallel cuts produced by conventional decision tree algorithms.

In principle an oblique decision tree could produce cuts of arbitrary complexity in data space, say quadratic or cubic cuts. In practice linear cuts are by far the most used, due mainly to their relative simplicity, and hereafter we assume that an oblique decision tree algorithm produces nodes with linear combinations of attributes.

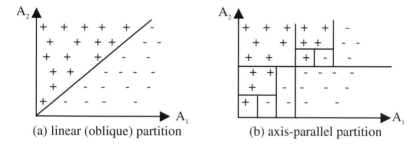

(a) linear (oblique) partition   (b) axis-parallel partition

Figure 3.3: Difference between axis-parallel and oblique data partitions

Oblique decision tree algorithms are more flexible than axis-parallel ones, since the latter is a special case of the former. Hence, oblique decision trees intuitively tend to achieve a higher classification accuracy in cases where the true classification function is better described by oblique hyperplanes than by axis-parallel ones. One example is given in Figure 3.3. The figure concerns a hypothetical data set with two predictor attributes, A_1 and A_2, and two classes, denoted by "+" and "-". An accurate separation between the classes can be easily obtained by the data partition of Figure 3.3(a), which corresponds to a decision tree with a single internal node having the linear combination: $(A_2 - A_1 > 0)$. If this condition is true for a given data instance it is classified as "+", otherwise it is classified as "-". In contrast, a conventional, axis-parallel decision tree would produce a more complex data partition, as shown in Figure 3.3(b).

However, the greater flexibility of oblique decision trees is achieved at the expense of some disadvantages. First, the search space associated with oblique decision trees is much larger than the search space associated with axis-parallel decision trees. (This can be considered a motivation for applying evolutionary algorithms in oblique decision-tree building, as will be seen in subsection 7.5.2.) As a result, oblique decision tree algorithms are usually much slower than their axis-parallel counterpart.

Another problem is that in an oblique decision tree each node can be a mathematical inequality, involving several attributes and corresponding coefficients. Most users consider this representation much harder to interpret, in comparison with the test based on a single attribute used by axis-parallel decision trees. This loss of comprehensibility is probably the main reason why oblique decision trees are much less popular than axis-parallel ones. Oblique-decision-tree building algorithms are described, e.g., in [Murthy et al. 1994] and [Wang and Zaniolo 2000].

3.2 Rule Induction Algorithms

A decision-tree building (or induction) algorithm can be considered a specific kind of rule induction algorithm, since there is a straightforward way of converting a decision tree into a rule set, as discussed above. However, in the literature the term rule induction algorithm is often used to refer to an algorithm which discovers a rule set somewhat more "flexible" than a decision tree, in the sense that the discovered (or induced) rules cover data space regions that can have some overlapping. Hence, a data instance can be covered by more than one rule. In contrast, recall that a decision tree produces rules representing data partitions that are mutually exclusive and collectively exhaustive. Hence, each data instance is covered by exactly one rule.

In other words, there is a certain difference in the kind of model induced by the two kinds of algorithm. Although in both cases the model is essentially a rule set, the rule set induced by rule induction algorithms can be considered more flexible than the rule set induced by decision-tree building algorithms. In particular, the latter can be considered a special case of the former where there is no overlapping between rules.

A general, high-level description of rule induction algorithms is shown in Algorithm 3.2. First of all, in the first two lines of Algorithm 3.2(a) the training set *T* is initialized with all training data instances and the discovered rule set (*Rule-Set*) is initialized with the empty set. Then the algorithm iteratively performs a loop where it calls an Induce-Rule procedure, adds the rule induced by this procedure into the current rule set and removes the data instances correctly covered by the just-induced rule from the current training set. By "covered instances" we mean the instances satisfying the antecedent (IF part) of the just-induced rule, whereas by "correctly covered" we mean that usually only the covered instances having the same class as the class predicted by the rule consequent (THEN part) are removed. This loop is repeated until a stopping criterion is satisfied, e.g., until all data instances are covered by at least one induced rule, or until the number of uncovered instances is smaller than a small pre-specified threshold.

One possible version of the Induce-Rule procedure is shown in Algorithm 3.2(b). This particular version constructs a rule in a top-down, general-to-specific fashion. More precisely, the procedure shown in Algorithm 3.2(b) starts with an empty induced rule, i.e., a rule with no condition. Next it iteratively performs a loop where it chooses the best condition to be added to the current induced rule, based on a given evaluation function and on the current contents of T, and then it adds the selected condition to the current induced rule. This loop is repeated while the induced rule can be improved by adding a new condition to it. At this point the current induced rule is returned to the procedure shown in Algorithm 3.2(a).

T = {all training data instances};
$RuleSet = \varnothing$;
while (stopping criterion
      is not satisfied)
  call Induce-Rule(T);
  $RuleSet = RuleSet \cup InducedRule$;
  $T = T - $ {instances covered
      by $InducedRule$};
endwhile

(a) Rule set construction
   (one-rule-at-a-time)

procedure Induce-Rule(T)
$InducedRule = \varnothing$;
while ($InducedRule$ can be improved
      by adding a new condition to it)
  choose $BestCondition$;
  $InducedRule = InducedRule \cup$
          $BestCondition$;
endwhile
return ($InducedRule$)

(b) Rule construction
   (one-condition-at-a-time)

Algorithm 3.2: General, high-level description of rule induction algorithms

---

It is important to notice that, despite the above-mentioned difference in the kind of model (rule set) induced by decision-tree building and rule induction algorithms, in general both kinds of algorithm perform a greedy search for rules. Similarly to the use of the term greedy in subsection 3.1.3, by greedy we mean these algorithms construct a rule in a step-by-step fashion, adding one-condition-at-a-time to the current partial rule, and at each step the best possible local choice is made. The main difference between the two kinds of algorithm is what constitutes a condition in their greedy search, as follows. Decision-tree building algorithms usually select one-*attribute*-at-a-time to be added to the current tree, as indicated in step (3) of Algorithm 3.1. In contrast, rule induction algorithms usually select one-*attribute-value-pair*-at-a-time to be added to the current rule, as shown in Algorithm 3.2(b) – the term *condition* used in the algorithm's description is a shorthand for an attribute-value pair, such as *Gender = male*.

3.2.1 The Pitfall of Attribute Interaction for Greedy Algorithms

We will illustrate the effect of attribute interaction on greedy rule induction algorithms with a very simple, pedagogical example, and then we will discuss more realistic scenarios in real-world data sets. In this subsection we also argue that evolutionary algorithms tend to cope better with attribute interaction than greedy rule induction algorithms.

Suppose a company is reviewing the records of its employees to detect possible irregularities in their payment. To keep the example very simple, let us suppose the company is trying to predict the value of the goal attribute *Irregular_payment?* using only three predictor attributes, namely the amount of tax paid by an employee (*Taxes*) and the employee's *Salary* and *Gender*. Finally, suppose a rule induction algorithm is given the very small data set shown in Table 3.2.

Table 3.2: Attribute interaction in a kind of logic XOR (eXclusive OR) function

Taxes	Salary	Gender	Irregular_payment?
low	low	male	no
low	high	male	yes
high	low	male	yes
high	high	female	no

Note that in this data set the attribute *Irregular_payment?* takes on the value *yes* if the value of *Taxes* is different from *Salary,* and it takes on the value *no* otherwise. In other words, the value of *Irregular_payment?* can be accurately predicted by a kind of logic XOR (eXclusive OR) function applied to attributes *Taxes* and *Salary*. The XOR function returns *yes* (*true*) if and only if one of the arguments is *high* (*true*) and the other one is *low* (*false*).

Unfortunately, in general a greedy rule induction procedure – like the one in Algorithm 3.2(b) – would fail in discovering such XOR-like relationship. To see why, let us see how this kind of algorithm would try to induce a rule from the data in Table 3.2 by selecting one attribute-value condition at a time. Note that the class distribution for the data set as a whole is 50%-50%, i.e., two data instances have class *yes* and two have class *no*.

Suppose the algorithm generates a candidate rule containing the antecedent: "IF (*Taxes = low*)". This rule antecedent is not helpful for predicting the value of *Irregular_payment?*, since the class distribution for the two examples satisfying this antecedent is 50%-50%, which is the same class distribution as in the entire data set. The same holds for all the other three possible rule antecedents that can be formed by using only one condition (attribute-value pair) referring to either *Taxes* or *Salary*, that is, the following three rule antecedents: "IF (*Taxes = high*)", "IF (*Salary= low*)", "IF (*Salary = high*)".

Therefore, a greedy rule induction algorithm, which selects rule conditions on an one-at-a-time basis, would consider that attributes *Taxes* and *Salary* are irrelevant for predicting *Irregular_payment?*. However, this is not the case. *Taxes* and *Salary* are relevant for predicting *Irregular_Payment?*. The problem is that we need to know the value of both these predictor attributes at the same time, to take into account their interaction, and in general this is not done by greedy rule induction algorithms.

As a result, this kind of algorithm would conclude that the attribute *Gender* is more relevant for predicting *Irregular_Payment?*, since there is an apparent (but in reality spurious) correlation between these two attributes. Hence, a greedy rule induction algorithm would be fooled and would unfortunately add a condition involving *Gender* – rather than *Taxes* or *Salary* – to a rule antecedent.

It should be noted that the above-discussed XOR-like attribute interaction problem is in reality a particular a case of parity problems, where the target function returns true if and only if an odd number of predictor attributes is true. The complexity of attribute interaction in parity problems increases very fast with the number of predictor attributes, which makes this kind of problem very difficult for greedy rule induction algorithms [Schaffer 1993].

We emphasize that the example of attribute interaction in the data set of Table 3.2 was oversimplified, since real-world data sets have much larger numbers of data instances and attributes. However, what is important in that example is not the details of the data set, but rather the concept of attribute interaction [Freitas 2001].

Attribute interaction problems do occur in real-world data sets, although in a much more subtle, hidden manner. For one thing, the above discussion focused on two-way attribute interaction – involving two predictor attributes – but higher-order attribute interactions are also possible.

A good example of evidence for the existence of attribute interaction problems in real-world data sets can be found in [Dhar et al. 2000]. In this work the authors discuss several characteristics of financial data sets which make prediction in this kind of application domain notoriously difficult. One of the characteristics is the existence of subtle, hidden attribute interactions which often are not detected by manual analysis nor by greedy rule induction algorithms. Quoting [Dhar et al. 2000, p. 251]:

"...in financial practice, where analysts conduct extensive manual analysis of historically well performing indicators, a key is to find the hidden interactions among variables that perform well in combination. Unfortunately, these are exactly the patterns that the greedy search biases incorporated by many standard rule learning algorithms will miss."

This work also reports the results of experiments comparing a genetic algorithm (GA) with two greedy rule induction algorithms, and presents some evidence that the GA is more effective at finding hidden attribute interactions. It should be stressed that the GA used in this work is not a simple, traditional GA, but rather a GA extended with several rule induction concepts and principles, i.e., a GA tailored for rule discovery.

Another evidence for the existence of attribute interaction problems in real-world application domains involves the problem of small disjuncts. To explain the meaning of this term, consider that we have a set of prediction rules in disjunctive normal form, i.e., a disjunction of rules where each rule is a conjunction of conditions. Then each rule represents a disjunct, and it can be considered as a small disjunct if it covers a small number of training data instances [Holte et al. 1989]. One of the major causes of the problem of small disjuncts is attribute interaction. When the data contains a strong degree of attribute interaction, in order to accurately predict the class of a data instance one needs to know the values of several predictor attributes at a time. This requires a conjunction of several conditions in a rule antecedent, which reduces the number of data instances covered by the rule, creating small disjuncts. The problem of small disjuncts occurs in several real-worlds data sets [Weiss and Hirsh 2000; Carvalho and Freitas 2002a, 2001]. For example, [Danyluk and Provost 1993] report a real-world application where small disjuncts cover roughly 50% of the data instances.

Note that it is possible to develop a hybrid method that combines the advantages of both greedy rule induction (or decision tree induction) and evolutionary algorithms. For instance, [Carvalho and Freitas 2000a, 2000b, 2002a] proposed a hybrid decision tree/genetic algorithm method. In essence, the method starts by running a conventional decision tree algorithm. The leaf nodes containing a large number of data instances are considered "large" (non-small) disjuncts. Instances belonging to those nodes are simply classified by the decision tree. In contrast, the leaf nodes containing a small number of instances are considered small disjuncts. For each of these leaf nodes a GA is run to discover rules that classify the instances belonging to that node. A more recent version of this hybrid method is proposed in [Carvalho & Freitas 2002b], where all the instances belonging to all the leaf nodes considered small disjuncts are grouped in a single training set, called the "second training set" (to distinguish it from the original training set used to build the decision tree). This second training set is provided as input data for a GA, in order to discover rules classifying all the small-disjunct instances. In both versions of this hybrid method, the basic idea is that large-disjunct instances, which are relatively easy to classify, are classified by a greedy decision tree algorithm; whereas small-disjunct instances, which are relatively difficult to classify (due to the higher amount of attribute interaction often associated with small disjuncts) are classified by a GA.

Furthermore, some authors have reported results showing that in general evolutionary algorithms outperform greedy rule induction algorithms in artificially-constructed data sets with a large degree of attribute interaction – see, e.g., [Papagelis and Kalles 2001; Greene and Smith 1993].

3.3 Instance-Based Learning (Nearest Neighbor) Algorithms

In both the decision-tree building and rule induction paradigms, discussed in the previous two sections, the goal was to induce an explicit classification model (a decision tree or a rule set) to classify new data instances. In contrast, in the instance-based learning (or nearest neighbor) paradigm the algorithm usually does not induce any explicit classification model. It simply uses the available data set, or a selected subset of it, to classify new data instances [Aha 1997; Aha et al. 1991; Aha 1992; Dasarathy 1991].

In essence, when a new data instance has to be classified the algorithm performs a matching between that instance and each of the stored instances (the training set). Then it retrieves the k training instances that are "nearest" (most similar) to the new instance (to be classified), where k, a parameter of the algorithm, is usually a small integer number. The new instance is then usually assigned the class of the majority of the k retrieved instances. This basic idea is illustrated in Figure 3.4, where MIN denotes a minimum operator that selects the k instances with smallest distance measure.

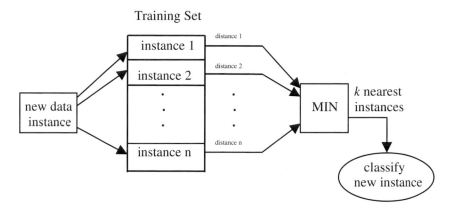

Figure 3.4: The basic idea of instance-based learning (nearest neighbor)

This kind of algorithm is sometimes called k-nn, a short for k nearest neighbors. The term nearest neighbor is typically used in the statistical pattern recognition literature, whereas the term instance-based learning is typically used in the machine learning literature. Since the induction process is delayed until classification, this paradigm is sometimes called "lazy" learning, to distinguish it from other paradigms, such as decision-tree building and rule induction, where the induction process is typically "eager". We prefer the terms instance-based learning or nearest neighbors, because the basic idea of lazy learning can be used in the decision-tree building and rule induction paradigms as well [Friedman et al. 1996; Fulton et al. 1996].

When computing the distance between the new instance (to be classified) and each training instance, it is common to use attribute weights to indicate the relevance of each attribute for classification. If an attribute is irrelevant it should have a small weight, so that differences in the values of that attribute will have a small influence on the value of the distance between the new instance and each training instance. Conversely, if an attribute is relevant it should have a large weight, so that differences in the values of that attribute will have a large influence on the value of the distance between the new instance and each training instance. Several methods for finding good attribute weights are discussed in [Wettschereck et al. 1997]. The search for good attribute weights can also be performed by an evolutionary algorithm, as will be discussed in sections 9.2 and 9.3.

In any case, it should be noted that any distance measure (with or without attribute weights) used to determine nearest neighbors is a form of inductive bias (section 2.5), so that it is suitable for some data sets and unsuitable for others [Wess and Globig 1993; Salzberg 1991].

Finally, note that although conventional nearest neighbor algorithms do not induce any explicit classification model, they *do* provide an explanation about how the classification of a new data instance is done, which is important in the context of data mining. When a new instance is classified, the system can show the user the classifying instances, i.e., the most similar training instances retrieved by the system. This kind of knowledge can be validated by the user, and it can be considered interpretable, as long as the number of nearest neighbors is small and the number of attributes is not large. In general the output of a nearest neighbor algorithm may not be so intuitively comprehensible as the output of a rule induction algorithm, but the former is more comprehensible than the output of a black box system.

References

[Aha 1992] D.W. Aha. Tolerating noisy, irrelevant and novel attributes in instance-based learning algorithms. *International J. Man-Machine Studies 36,* 267–287, 1992.

[Aha 1997] D.W. Aha (Ed.) *Artificial Intelligence Review – Special issue on lazy learning, 11(1–5),* June 1997.

[Aha et al. 1991] D.W. Aha, D. Kibler and M.K. Albert. Instance-based learning algorithms. *Machine Learning 6,* 37–66, 1991.

[Bohanec and Bratko 1994] M. Bohanec and I. Bratko. Trading accuracy for simplicity in decision trees. *Machine Learning 15,* 223–250, 1994.

[Bramer 1996] M. Bramer. Induction of classification rules from examples: a critical review. *Proceedings of the Data Mining'96 Unicom Seminar,* 139–166. London: Unicom, 1996.

[Breiman et al. 1984] L. Breiman, J.H. Friedman, R.A. Olshen and C.J. Stone. *Classification and Regression Trees.* Pacific Groves, CA: Wadsworth, 1984.

[Breslow and Aha 1997] L.A. Breslow and D.W. Aha. Simplifying decision trees: a survey. *The Knowledge Engineering Review 12(1),* 1–40, Mar. 1997.

[Carvalho and Freitas 2000a] D.R. Carvalho and A.A. Freitas. A hybrid decision tree/genetic algorithm for coping with the problem of small disjuncts in data mining. *Proceedings of the Genetic and Evolutionary Computation Conference (GECCO '2000),* 1061–1068. Morgan Kaufmann, 2000.

[Carvalho and Freitas 2000b] D.R. Carvalho and A.A. Freitas. A genetic algorithm-based solution for the problem of small disjuncts. *Principles of Data Mining and Knowledge Discovery (Proceedings of the 4th European Conference, PKDD '2000). Lecture Notes in Artificial Intelligence 1910,* 345–352. Springer, 2000.

[Carvalho and Freitas 2001] D.R. Carvalho and A.A. Freitas. An immunological algorithm for discovering small-disjunct rules in data mining. *Proceedings of the Graduate Student Workshop held at the Genetic and Evolutionary Computation Conference (GECCO '2001),* 401–404. San Francisco, CA, USA. 2001.

[Carvalho and Freitas 2002a] D.R. Carvalho and A.A. Freitas. A genetic algorithm for discovering small-disjunct rules in data mining. To appear in *Applied Soft Computing journal.* 2002.

[Carvalho and Freitas 2002b] D.R. Carvalho and A.A. Freitas. A genetic algorithm with sequential niching for discovering small-disjunct rules. To appear in *Proceedings of the Genetic and Evolutionary Computation Conference (GECCO '2002)*. New York, 2002.

[Catlett 1991] J. Catlett. Overpruning large decision trees. *Proceedings of the 12th International Joint Conference AI (IJCAI '91)*. Sidney, 1991.

[Cendrowska 1987] J. Cendrowska. PRISM: an algorithm for inducing modular rules. *International Journal of Man-Machine Studies*, 27, 349–370, 1987.

[Danyluk and Provost 1993] A.P. Danyluk and F.J. Provost. Small disjuncts in action: learning to diagnose errors in the local loop of the telephone network. *Proceedings of the 10th International Conference on Machine Learning (ICML '93)*, 81–88. Morgan Kaufmann, 1993.

[Dasarathy 1991] B.V. Dasarathy (Ed.) *Nearest Neighbor (NN) Norms: NN Pattern Classification Techniques*. IEEE Computer Society Press, 1991.

[Dhar and Stein 1997] V. Dhar and R. Stein. *Seven Methods for Transforming Corporate Data into Business Intelligence*. Prentice-Hall, 1997.

[Dhar et al. 2000] V. Dhar, D. Chou and F. Provost. Discovering interesting patterns for investment decision making with GLOWER – a genetic learner overlaid with entropy reduction. *Data Mining and Knowledge Discovery 4(4)*, 251–280. 2000.

[Esposito et al. 1993] F. Esposito, D. Malerba and G. Semeraro. Decision tree pruning as search in the state space. *Machine Learning: ECML '93. LNAI-667*, 165–184. Springer, 1993.

[Fayyad 1994] U.M. Fayyad. Branching on attribute values in decision tree generation. *Proceedings of the 12th National Conference on Artificial Intelligence (AAAI '94)*, 601–606. AAAI Press, 1994.

[Fayyad and Irani 1993] U.M. Fayyad and K.B. Irani. Multi-interval discretization of continuous-valued attributes for classification learning. *Proceedings of the 13th International Joint Conference Artificial Intelligence*, 1022–1027. Chamberry, France, Aug./ Sep. 1993.

[Fayyad et al. 1993] U.M. Fayyad, N. Weir and S. Djorgovski. SKICAT: a machine learning system for automated cataloging of large scale sky surveys. *Proceedings of the 10th International Conference Machine Learning*, 112–119. Morgan Kaufmann, 1993.

[Fisher and Hapanyengwi 1993] D. Fisher and G. Hapanyengwi. Database management and analysis tools of machine induction. *Journal of Intelligent Information Systems*, 2(1), 5–38, Mar. 1993.

[Freitas 2001] A.A. Freitas. Understanding the crucial role of attribute interaction in data mining. *Artificial Intelligence Review 16(3)*, 177–199, 2001.

[Freitas and Lavington 1998] A.A. Freitas and S.H. Lavington. *Mining Very Large Databases with Parallel Processing*. Kluwer, 1998.

[Friedman et al. 1996] J.H. Friedman, R. Kohavi and Y. Yun. Lazy decision trees. *Proceedings of the 14th National Conference of the American Association for Artificial Intelligence (AAAI '96)*, 1996.

[Fulton et al. 96] T. Fulton, S. Kasif, S. Salzberg and D. Waltz. Local induction of decision trees: towards interactive data mining. *Proceedings of the 2nd International Conference Knowledge Discovery and Data Mining (KDD '96)*, 14–19. AAAI Press, 1996.

[Gehrke et al. 2000] J. Gehrke, R. Ramakrishnan and V. Ganti. RainForest – A framework for fast decision tree construction of large data sets. *Data Mining and Knowledge Discovery journal 4(2/3)*, 127–162, 2000.

[Greene and Smith 1993] D.P. Greene and S.F. Smith. Competition-based induction of decision models from examples. *Machine Learning 13*, 229–257, 1993.

[Holte et al. 1989] R.C. Holte, L.E. Acker and B.W. Porter. Concept learning and the problem of small disjuncts. *Proceedings of the 1989 International Joint Conference on Artificial Intelligence (IJCAI '89)*, 813–818. 1989

[Langley 1996] P. Langley. *Elements of Machine Learning*. Morgan Kaufmann, 1996.

[Lavington et al. 1999] S. Lavington, N. Dewhurst, E. Wilkins, and A.A. Freitas. Interfacing knowledge discovery algorithms to large database management systems. *Information and Software Technology 41 (1999), special issue on data mining*, 605–617. 1999.

[Lim et al. 2000] T.-S. Lim, W.-Y. Loh and Y.-W. Shih. A comparison of prediction accuracy, complexity, and training time of thirty-three old and new classification algorithms. *Machine Learning 40*, 203–228, 2000.

[Michie et al. 1994] D. Michie, D.J. Spiegelhalter and C.C. Taylor. *Machine Learning, Neural and Statistical Classification*. Ellis Horwood, 1994.

[Mingers 1989a] J. Mingers. An empirical comparison of pruning methods for decision tree induction. *Machine Learning 4*, 227–243, 1989.

[Mingers 1989b] J. Mingers. An empirical comparison of selection measures for decision-tree induction. *Machine Learning 3*, 319–342, 1989.

[Mitchell 1997] T. Mitchell. *Machine Learning*. McGraw-Hill, 1997.

[Murthy 1998] S.K. Murthy. Automatic construction of decision-trees from data: a multi-disciplinary survey. *Data Mining and Knowledge Discovery journal, 2(4)*, 345–389. 1998.

[Murthy et al. 1994] S.K. Murthy, S. Kasif and S. Salzberg. A system for induction of oblique decision trees. *Journal of Artificial Intelligent Research 2*, 1–32, 1994.

[Pagallo and Haussler 1990] G. Pagallo and D. Haussler. Boolean feature discovery in empirical learning. *Machine Learning 5*, 71–99, 1990.

[Papagelis and Kalles 2001] A. Papagelis and D. Kalles. Breeding decision trees using evolutionary techniques. *Proceedings of the 18th International Conference on Machine Learning (ICML '2001)*, 393–400. San Francisco, CA: Morgan Kaufmann, 2001.

[Quinlan 1987a] J.R. Quinlan. Generating production rules from decision trees. *Proceedings of the International Joint Conference on Artificial Intelligence (IJCAI '87)*, 304–307, 1987.

[Quinlan 1987b] J.R. Quinlan. Simplifying decision trees. *International Journal of Man-Machine Studies 27*, 221–234, 1987.

[Quinlan 1993] J.R. Quinlan. *C4.5: Programs for Machine Learning*. Morgan Kaufmann, San Mateo, 1993.

[Quinlan 1996] J.R. Quinlan. Improved use of continuous attributes in C4.5. *J. of AI Research (JAIR) 4*, 77–90, 1996.

[Salzberg 1991] S. Salzberg. Distance metrics for instance-based learning. *Proceedings of the 6th International Symposium on Methodologies for Intelligent*

Systems (ISMIS' 91). Lecture Notes in Artificial Intelligence 542, 399–408. Springer, 1991.

[Schaffer 1993] C. Schaffer. Overfitting avoidance as bias. *Machine Learning* 10, 153–178, 1993.

[Srivastava et al. 1999] A. Srivastava, E-H. Han, V. Kumar and V. Singh. Parallel formulations of decision-tree classification algorithms. *Data Mining and Knowledge Discovery journal 3(3)*, 237–261, Sep. 1999.

[Wang and Zaniolo 2000] H. Wang and C. Zaniolo. CMP: a fast decision tree classifier using multivariate predictions. *Proceedings of the International Conference on Data Engineering* 2000.

[Weiss and Hirsh 2000] G.M. Weiss and H. Hirsh. A quantitative study of small disjuncts. *Proceedings of the 17th National Conference on Artificial Intelligence (ΛΛΛΙ '2000)*, 665–670. AAAI Press, 2000.

[Wess and Globig 93] S. Wess and C. Globig. Case-based and symbolic classification: a case study. *Proceedings of the 1st European Workshop on CBR (EWCBR' 93). Lecture Notes in Artificial Intelligence 837*, 77–91. Springer, 1994.

[Wettschereck et al. 1997] D. Wettschereck and T.G. Dietterich. An experimental comparison of the nearest-neighbor and nearest-hyperrectangle algorithms. *Machine Learning* 19, 5–27, 1995.

[White and Liu 1994] A.P. White and W.Z. Liu. Bias in information-based measures in decision tree induction. *Machine Learning 15,* 321–329, 1994.

4 Data Preparation

> "Garbage in, garbage out"
> (old Computer Science saying)

No matter how "intelligent" a data mining algorithm is, it will fail to discover high-quality knowledge if it is applied to low-quality data. In this chapter we focus on data preparation methods for data mining. The general goal is to improve the quality of the data being mined, to facilitate the application of a data mining algorithm. Hence, the methods discussed in this chapter can be regarded as a form of preprocessing for a data mining algorithm.

Section 4.1 discusses the problem of attribute selection, also called feature selection. Section 4.2 discusses the discretization of continuous attributes. Finally, section 4.3 discusses the problem of attribute construction, also called feature construction or constructive induction. The contents of this chapter is particularly useful for a better understanding of chapter 9, where we will discuss evolutionary algorithms for data preparation.

4.1 Attribute Selection

In this section we are interested in attribute selection as a data preparation task for data mining. The main goal of attribute selection is to select a subset of relevant attributes, out of all available attributes of the data being mined [John et al. 1994; Kohavi and John 1998]. Then the selected attributes will be given to a data mining algorithm. Note that the number of candidate attribute subsets (the size of the search space of the attribute selection task) is 2^m, where m is the number of attributes. Hence, the size of the search space grows exponentially with the number of attributes.

In general the kind of data selection method to be used obviously depends on the data mining task. Throughout this section we will assume – unless mentioned otherwise – that the data mining task being addressed is classification, since this is by far the task most investigated in the attribute selection literature. However, the basic ideas of this section also hold in the case of other data mining tasks involving prediction, such as the dependence modeling task (section 2.2).

The remainder of this section is divided into two parts, as follows. Subsection 4.1.1 reviews the motivation for attribute selection as a preprocessing task for data mining. Subsection 4.1.2 presents an overview of the filter and wrapper approaches for attribute selection. A more comprehensive discussion on attribute selection can be found in [Liu and Motoda 1998].

4.1.1   The Motivation for Attribute Selection

Attribute selection is a very important preprocessing task in data mining, particularly in the discovery of prediction rules. After all, if we want to discover rules predicting the value of a given goal attribute, it is crucial that the predictor attributes given to rule-discovery algorithms be relevant for predicting that goal attribute.

Let us consider a very simple example of the importance of attribute selection. Suppose that we want to predict whether or not a patient will develop a given disease, based on a number of predictor attributes about the patient. Presumably, our patient database contains a unique id number for each patient. If this attribute is included in the set of attributes given to the rule discovery algorithm, it is possible that the algorithm will discover rules of the form:

IF (*Unique_Id = 1234*) THEN (*Disease* = yes).

Although this rule is correct in the training data it is useless for predictions on an unseen test set. It is too specific, and it overfits the training data. Hence, attributes with no generalization power – such as *Unique_Id* in the previous example – should be removed from the set of predictor attributes given to the rule discovery algorithm.

There are several possible motivations for attribute selection [John et al. 1994; Koller and Sahami 1996]. Firstly, a well-selected subset of attributes often leads to the discovery of rules with higher predictive accuracy than the entire set of attributes. This effect is sometimes informally described as "less is more" in the literature. The basic idea is that irrelevant attributes can deceive or "confuse" the data mining algorithm, so that removing those attributes can actually improve the quality of discovered rules.

Secondly, *if* the attribute selection method is fast, the total time taken to run the attribute selection method and then run the classification algorithm using only the selected attributes might be shorter than the time taken to run the classification algorithm using the entire set of attributes. In practice, however, an attribute selection method can be quite slow, so that attribute selection can actually increase the total processing time. In particular, this is usually the case when using the wrapper approach, to be discussed later.

Thirdly, the selected attribute subset can lead to simpler (shorter) discovered rules, by comparison with the full set of original attributes. Such rule simplicity is often desired in the context of data mining.

4.1.2   Filter and Wrapper Approaches

Attribute selection methods can be categorized into two broad approaches: the filter approach and the wrapper approach. The key difference between these two approaches is as follows. In the filter approach attribute selection is performed without taking into account the classification algorithm that will be applied to the selected attributes. In this approach the goal is to select a subset of attributes that preserves as much as possible the relevant information found in the entire set of attributes.

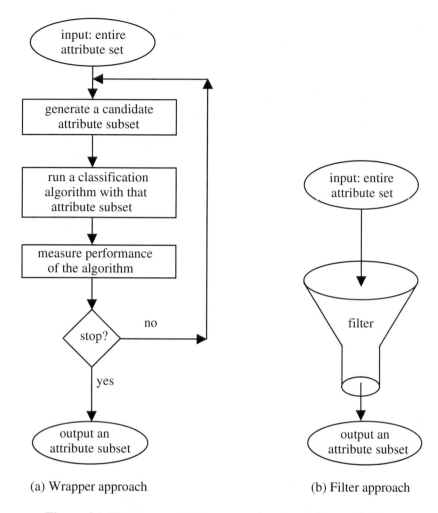

Figure 4.1: Wrapper and Filter approaches for attribute selection

In contrast, in the wrapper approach attribute selection is performed by taking into account the classification algorithm that will be applied to the selected attributes. In this approach the goal is to select a subset of attributes that is "optimized" for a given classification algorithm. This difference is illustrated in Figure 4.1. Note that in the wrapper approach the classification algorithm is run many times, each time with a different subset of attributes. The performance of the classification algorithm in each run is used to evaluate the quality of the corresponding attribute set. Note also that in this approach the classification algorithm is used as a black box, i.e., the attribute-selection method uses only the output of the classification algorithm, without any knowledge of its internal operations.

One example of a filter method for attribute selection can be found in [Wang and Sundaresh 1998]. The basic idea of this method is to use a vertical compactness criterion for evaluating the quality of a given candidate attribute subset. The method starts by assuming that one has some idea about how much inconsistency can be tolerated in the data being mined. The term inconsistency refers to the situation where two or more data instances have the same value for all selected predictor attributes but have different goal-attribute values (classes). Once a maximum inconsistency rate is specified, the method searches for an attribute subset that produces the minimum number of projected instances, provided that the resulting inconsistency rate is not greater than the specified maximum. The "projected instances" of a given attribute subset S are the instances produced by eliminating all attributes not specified in S and then eliminating duplicate instances from what is left.

To see the logic behind this criterion, let us consider two extreme cases. On one extreme, consider the attribute *Gender*, which can take on only the values *male* and *female*. This attribute produces only two projected instances. Unfortunately, in most real-world applications *Gender* alone will be associated with a very high inconsistency rate, since there will be probably many instances with the same value of *Gender* but different classes. On the other extreme, consider the above-mentioned attribute *Unique_Id*. This attribute alone will have a zero inconsistency rate, since there will be only one class associated with each value of this attribute. Unfortunately, this attribute has the highest possible number of projected instances, i.e., the number of instances being mined. This attribute has no generalization power.

To summarize, the challenge is to find an attribute subset that produces the least possible number of projected instances, provided that the resulting inconsistency rate is not greater than a specified maximum.

Let us now turn to the wrapper approach. One important point about this approach is that the performance of the classification algorithm – which is used to evaluate the quality of the current attribute subset – *cannot* be evaluated on the test set. If we did that we would be committing the "fatal sin" of using the test set for training. To avoid this one must reserve a part of the training set for the purpose of evaluating the performance of the classification algorithm within the loop of the attribute selection procedure. One simple way of doing this consists of randomly dividing the training set into two subsets, which we will call the "training-training" set and the "training-test" set. The former is used to train the classification algorithm. Once the algorithm is trained, its performance will be measured on the "training-test" set. This data set plays the role of an unseen test set for the classification algorithm within the loop of the wrapper approach, but it is included in the training set that will be used by the classification algorithm after the attribute selection method is run. Indeed, when the attribute-selection method terminates, the best attribute subset found by that method is given to the classification algorithm, which is then finally run on the entire training set. The knowledge discovered by the classification algorithm is then evaluated on the test set, whose data instances remained *unseen* during the entire run of the attribute-selection method.

So far our discussion has focused on how to use a classification algorithm for evaluating the quality of a candidate attribute subset. We now turn to the problem of how to generate candidate attribute subsets to be evaluated. Clearly, if the data being mined has a small number of attributes, we can simply generate all possible attribute subsets and measure the performance of the classification algorithm on each of these subsets. Unfortunately, the number of candidate attribute subsets grows exponentially with the number of available attributes, as mentioned above. In this section we are interested in non-trivial cases where the number of candidate attribute subsets is large enough to render it impractical the exhaustive search approach of generating and evaluating every possible attribute subset. In such cases one must perform some kind of heuristic search in the space of all possible candidate attribute subsets.

Two simple search strategies used in the wrapper approach for attribute selection are forward sequential selection and backward sequential elimination. Forward sequential selection starts with a candidate attribute subset that is the empty set. Then it iteratively adds one attribute at a time to the candidate attribute subset, as long as some measure of performance – say predictive accuracy – is improved. At each iteration the attribute added to the current attribute subset is the attribute whose inclusion maximizes the measure of performance.

Backward sequential elimination is the opposite strategy. It starts with a candidate attribute subset that is the entire set of available attributes. Then it iteratively removes one attribute at a time from the candidate attribute subset, as long as some measure of performance is improved. At each iteration the attribute removed from the current attribute subset is the attribute whose removal maximizes the measure of performance.

Note that both forward selection and backward elimination are greedy search strategies, which can be trapped in local maxima in the search space. To reduce this possibility one can use more robust search strategies. In particular, the use of genetic algorithms for attribute selection, following the wrapper approach, will be discussed in detail in section 9.1.

Finally, let us say a few words about the measure of performance of the classification algorithm. This measure must be carefully chosen, since in the wrapper approach it is used to guide the search for a good subset of candidate attributes. In many cases the chosen measure is simply the predictive accuracy of the classification algorithm on the training-test data subset. However, it is important to bear in mind that in data mining the discovered rules should be not only accurate, but also comprehensible and interesting for the user (see subsection 1.1.1 and section 2.3). Therefore, it makes sense to define a classification-algorithm performance measure that takes into account these additional criteria of rule quality.

4.1.2.1 Pros and Cons of Filter and Wrapper Approaches

Let us now review the pros and cons of the filter and wrapper approaches. The wrapper approach selects an attribute subset that is optimized for a given classification algorithm, whereas the filter approach selects an attribute subset independent of the classification algorithm. Therefore, one intuitively expects that, for a given classification algorithm, the wrapper approach tends to lead to a higher predictive accuracy than the filter approach.

On the other hand, in general the wrapper approach is much slower than the filter approach, since in the former we have to run a full classification algorithm on many different attribute subsets. In addition, the wrapper approach implies a certain loss of generality, since the selected attributes are optimized for just one classification algorithm. In contrast, in the filter approach the attribute selection method is independent of any particular classification algorithm, so that the selected attributes are supposed to be useful for several different classification algorithms.

Given this trade-off between predictive accuracy and processing time, it is interesting to ask how we could compare the *cost-effectiveness* of both approaches – wrapper and filter. A first step towards answering this question was suggested by [Freitas 1997], based on the simple idea that a reduction in processing time can be "transformed" into a gain in predictive accuracy.

In the context of attribute selection, this idea can be explained as follows. Suppose we apply the wrapper approach with a given classification algorithm X to a given data set, and as a result we obtain a classification error rate $Error_W$ with a processing time of $Time_W$. In addition, suppose we apply the filter approach to the same data set, and when we apply the same classification algorithm X to the selected attributes we obtain an error rate $Error_F$ with a processing time of $Time_F$. As mentioned above, we expect $Error_W < Error_F$ but $Time_W >> Time_F$ (where "x >> y" denotes that x is much larger than y). Hence, the saving in processing time due to the use of the filter approach is given by $Time_W - Time_F$. Instead of using a wrapper approach, we can spend this saved time by applying several different classification algorithms to the attributes selected by the filter method. After we are done, we select as the best algorithm the algorithm which led to the smallest error rate among all algorithms tried (including X). Since the idea of applying several classification algorithms to a data set and select the best one is often called a toolbox approach in the literature, we call the whole above approach a filter/toolbox approach.

It is possible that this filter/toolbox approach be more cost-effective than the wrapper approach, in the sense that if we try both approaches, by allocating a roughly equal amount of time to both of them, the filter/toolbox approach can lead to a better predictive accuracy. After all, as mentioned in section 2.5, it has been shown, both theoretically and empirically, that the error rate of a classification algorithm is strongly dependent on the data set being mined. In practice it is difficult to decide *a priori* which is the "best" classification algorithm for the target data set, which is a crucial decision in the wrapper approach. The filter/toolbox approach avoids this difficult decision by simply trying several different algorithms on the target data set and picking the best-performance one.

In any case, we are not aware of any experiments comparing the cost-effectiveness of the filter/toolbox and wrapper approaches in the above described fashion. We believe that an extensive set of experiments comparing these two approaches, involving several data sets, could produce interesting results for data mining researchers and practitioners.

4.1.3 Attribute Selection as a Particular Case of Attribute Weighting

In attribute weighting the goal is to assign to each attribute a numerical weight that indicates the relevance of the attribute for classification. The higher the relevance of the attribute, the higher its weight should be. In other words, the goal is to search for a relative weighting of attributes that gives optimal classification performance. Of course, this assumes that the classification algorithm somehow uses attribute weights, which is typically the case in the nearest neighbor paradigm (section 3.3).

Attribute selection can be regarded as a particular case of attribute weighting. To see why this is the case assume, without loss of generality, that the attribute weights vary in the range [0..1]. (If another range is used, one can always normalize the attribute weights so that they fall into the range [0..1].) In attribute selection we can assign just two "weights" to each attribute – either 0, indicating that the attribute is not selected, or 1, indicating that the attribute is selected – while in attribute weighting the weights can take on any value in the [0..1] continuum.

4.2 Discretization of Continuous Attributes

As a form of data preparation for data mining, continuous (real-valued) attributes can be discretized, i.e., transformed into discrete attributes. The discretized attributes can then be treated as categorical attributes, taking on a small number of values. The basic idea of discretization is to partition the values of a continuous attribute into a small list of intervals. Each resulting interval is regarded as a discrete value of that attribute. This basic idea is illustrated in Figure 4.2, which shows a possible discretization for the attribute *Age*. The bottom of this figure shows an ordered list of continuous *Age* values. In this figure these continuous values were discretized into four intervals, shown at the top of the figure.

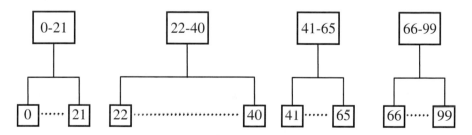

Figure 4.2: Example of discretization of attribute *Age*

4.2.1 An Overview of Discretization Methods

There are several kinds of discretization methods. One of the simplest ones is the equal-width method, which simply divides the range of values of a continuous attribute into a pre-specified number of equal-width intervals. For instance, this method could discretize the attribute *Age* into a list of intervals such as *0-20, 20-40, 40-60, 60-80* and *80-100*.

In the context of data mining tasks involving prediction, such as classification, we are usually interested in more sophisticated discretization methods. Actually, in the context of classification, discretization methods can be divided into two broad groups: class-blind and class-driven methods. Class-blind methods discretize an attribute without considering the classes (goal-attribute values) of data instances [Mannino et al. 88]. The above equal-width method is a typical example.

In contrast, class-driven methods consider the classes of data instances when discretizing an attribute. The input for this kind of method is usually a reduced data set with just two attributes: the attribute to be discretized and a goal attribute, whose value indicates the class to which the corresponding data instance belongs. Intuitively, when the intervals produced by a discretization method will be used by a classification algorithm, class-driven methods tend to produce better results than class-blind methods. This can be understood with a simple example, adapted from [Kerber 1992], as follows.

Suppose that the minimum *Age* required for a person to have a driver's license is *18*. In an application involving, say, the selling of car insurance, a class-driven discretization method would have the information necessary to identify the *Age* of *18* as the boundary of two intervals. In contrast, a class-blind method, say the equal-width method, could well produce an *Age* interval such as *0-20*. In this case no matter how intelligent the classification algorithm is, it would not be able to produce rules referring to the *Age* value of *18*.

Most class-driven discretization methods are essentially iterative algorithms that can be cast as a heuristic search in the space of possible discretizations. In each iteration the algorithm considers several candidate discretizations – different lower and upper values for intervals – and chooses the best one according to some candidate-discretization quality measure. This iterative procedure is performed until some stopping criterion is satisfied.

This procedure is usually applied to discretize one continuous attribute at a time. This has the disadvantage of ignoring attribute interactions, but it is still the most used approach for reasons of computational efficiency. An exception is the work of [Parthasarathy et al. 1998], where a discretization method considers two attributes at a time.

Several class-driven discretization methods are described in [Catlett 1991; Kerber 1992; Fayyad and Irani 1993; Richeldi and Rossotto 1995; Pfahringer 1995; Kohavi and Sahami 1996; Freitas and Lavington 1996; Ho and Scott 1997]. A review of several discretization algorithms is presented in [Dougherty et al. 1995; Hussain et al. 1999].

4.2.2 The Pros and Cons of Discretization

The pros and cons of discretization, as a preprocessing step for data mining, can be summarized as follows [Freitas and Lavington 1998, 38-39]. First of all, some rule-discovery algorithms cannot cope with continuous attributes. The use of this kind of algorithm requires that all data be discretized in a preprocessing step. The set of rule-discovery algorithms that cannot cope with continuous attributes includes both algorithms belonging to the rule induction paradigm, such as the algorithms described by [Apte and Hong 1996; Smyth and Goodman 1991], and algorithms belonging to the evolutionary computation paradigm, such as some algorithms discussed in chapter 6 of this book.

Discretization also has the advantage that it usually improves the comprehensibility of discovered rules [Catlett 1991; Pfahringer 1995]. Actually, discretization can be regarded as a data abstraction process, transforming lower-level data into higher-level data, as illustrated in Figure 4.2.

Another advantage of discretization is that it significantly reduces the time taken to run a classification algorithm belonging to the rule induction paradigm [Catlett 1991; Paliouras and Bree 1995; Freitas and Lavington 1996]. This is due to the fact that a discretization algorithm usually needs to sort the values of the continuous attribute being discretized just once, at the beginning of the algorithm. In contrast, many rule induction algorithms need to sort those values not only at the beginning of the algorithm, but also at a number of other moments during the run of the algorithm. As an extreme, but real-world example, [Catlett 1991] reported that applying a decision-tree algorithm to discretized data achieved a speed up (over applying the same algorithm to raw, non-discretized data) of more than 50 in one data set with 92 real-valued attributes. Such a large speed up was obtained without any significant decrease in classification accuracy. However, this advantage is not usually applicable to evolutionary algorithms, since this kind of algorithm does not usually sort continuous attribute values during the run of the algorithm.

At first glance, a potential disadvantage of discretization is that it may reduce the accuracy of the discovered knowledge. Intuitively, this might occur because discretization, similarly to any kind of data abstraction process, may cause some relevant detailed information to be lost. On the other hand, however, a discretization algorithm can make global decisions, based on all the values of the attribute to be discretized. This is in contrast with the local handling of continuous attributes' values typically found in an algorithm belonging to the rule induction paradigm. In practice, discretization may either increase or decrease classification accuracy, depending on the original data and on the data mining algorithm applied to the discretized data. Several studies where discretization improved classification accuracy in most cases have been reported in the literature [Kohavi and Sahami 1996; Richeldi and Rossotto 1995; Dougherty et al. 1995].

4.3 Attribute Construction

Attribute construction, also called feature construction or constructive induction, can be regarded as another data preparation task for data mining. In essence, the goal of attribute construction is to automatically construct new attributes, applying some operations/functions to the original attributes, so that the new attributes make the data mining problem easier [Rendell and Seshu 1990; Bloedorn and Michalski 1998; Hu 1998].

The motivation for this task is particularly strong when the original attributes do not have much predictive power by themselves. Intuitively, this situation tends to be relatively common in real-world database systems, since in general the data stored in these systems was not collected for data mining purposes. In these cases predictive accuracy can often be improved by constructing new attributes with greater predictive power.

To illustrate the importance of attribute construction, suppose that we want to discover rules for classifying the performance of a company into either *good* or *bad*, mining a very simple data set with the predictor attributes *Income* and *Expenditure*, as shown in Figure 4.3(a). Among the classification algorithms that discover knowledge expressed in the form of IF-THEN prediction rules, the majority of them discovers rules whose conditions (in the IF part of the rule) are essentially attribute-value pairs, such as *Income > 1000*. From a logical viewpoint, these rule conditions can be said to be expressed in propositional logic, or 0-th order logic – see section 1.2.1. This kind of rule representation is not useful for the data set shown in Figure 4.3(a). Why? Because what is relevant for predicting the performance of a company in this data set is not the *absolute* value of the attributes *Income* and *Expenditure* for a data instance, but rather the *relationship* between values of these attributes for a data instance. Unfortunately, most rule discovery algorithms cannot discover the rules:

IF (*Income > Expenditure*) THEN (*Performance = good*)
IF (*Income < Expenditure*) THEN (*Performance = bad*)

because these rules involve a first-order logic condition, namely the comparison between two attributes.

Now, suppose that we apply an attribute construction method to the data set of Figure 4.3(a), as a preprocessing step for the application of a classification algorithm. The attribute construction method could automatically construct a binary attribute such as "*Income>Expenditure?*", taking on the values "yes" or "no". The new attribute would then be given (together with the original attributes) to the classification algorithm, in order to improve the quality of the rules to be discovered by the latter. This corresponds to extending the original data set of Figure 4.3(a) with the new constructed attribute, as shown in Figure 4.3(b). Now it is easy for a propositional-logic rule discovery algorithm to discover the rules:

IF ("*Income>Expenditure?*" = *yes*) THEN (*Performance = good*)
IF ("*Income>Expenditure?*" = *no*) THEN (*Performance = bad*)

Income	Expend.	Perf.
900	750	good
730	610	good
510	400	good
620	390	good
340	450	bad
690	810	bad
450	590	bad
530	720	bad

Income	Expend.	Inc>Exp?	Perf.
900	750	yes	good
730	610	yes	good
510	400	yes	good
620	390	yes	good
340	450	no	bad
690	810	no	bad
450	590	no	bad
530	720	no	bad

(a) Original data set, with
just two predictor attributes

(b) Data set extended with
new attribute "Inc>Exp?"

Figure 4.3: An example of how attribute construction facilitates data mining

The major problem in attribute construction is that the search space tends to be huge. In the above very simple example it was easy to see that a good attribute could be constructed by using the comparison operator ">". However, in most real-world applications there will be a large number of attributes, as well as a large number of candidate operations to be applied to the original attributes, and it is far from clear which of those operations should be used (and on which original attributes). In addition, the construction of a good attribute often requires that the attribute construction algorithm generates and evaluates combinations of several original attributes, rather than just two attributes as in the above example. Therefore, the challenge is to design a cost-effective attribute construction method that constructs good-quality attributes in a reasonable processing time.

4.3.1 Categorizing Attribute Construction Methods

Attribute construction methods can be roughly categorized according to several dimensions [Hu 1998,   258-260]. Here we discuss the two dimensions that are probably the most used in the literature.

One dimension concerns the distinction between preprocessing and interleaving. In the preprocessing approach the attribute construction method is used as a preprocessor to any data mining algorithm. This has the advantage of generality, since the preprocessing method is separated from the data mining algorithm. In contrast, in the interleaving approach the attribute construction method and the data mining algorithm are intertwined into a single algorithm. This has the advantage that the constructed attributes are "optimized" for the data mining algorithm in question, but it has the disadvantage of losing generality. This approach can be implemented via the following iterative process. The attribute construction method produces new attributes that are given to the target data mining algorithm.

Then the data mining algorithm is run on a data set extended with the new attributes. The performance of the data mining algorithm is then fedback to the attribute construction method. This process is repeated until the data mining algorithm obtains a satisfactory predictive accuracy or until another stopping criterion is satisfied.

Another dimension for categorizing attribute selection methods concerns the distinction between hypothesis-driven and data-driven methods. Hypothesis-driven methods construct new attributes based on previously-generated "hypotheses". By hypotheses one means pieces of knowledge used to "explain the data". For instance, a decision tree or a rule set discovered by a data mining can be considered as hypotheses to explain the data being mined. As pointed out by Hu, hypothesis-driven methods have the advantage of using previous knowledge, but have the disadvantage of depending on the quality of the previous knowledge. In contrast, data-driven methods construct new attributes based on the data being mined. In other words, data-driven methods directly generate new attributes, by applying several functions to the original attributes. The new attributes are then evaluated according to a given attribute-quality measure.

References

[Apte and Hong 1996] C. Apte and S.J. Hong. Predicting equity returns from securities data. In: U.M. Fayyad, G. Piatetsky-Shapiro, P. Smyth and R. Uthurusamy (Eds.) *Advances in Knowledge Discovery and Data Mining,* 541–560. AAAI/MIT Press, 1996.

[Bloedorn and Michalski 1998] E. Bloedorn and R.S. Michalski. Data driven constructive induction: methodology and applications. In: H. Liu and H. Motoda (Eds.) *Feature Extraction, Construction and Selection: a data mining perspective,* 51–68. Kluwer, 1998.

[Catlett 1991] J. Catlett. On changing continuous attributes into ordered discrete attributes. *Proceedings of the European Working Session on Learning (EWSL-91). LNAI 482,* 164–178. Springer, 1991.

[Dougherty et al. 1995] J. Dougherty, R. Kohavi and M. Sahami. Supervised and unsupervised discretization of continuous features. *Proceedings of the 12th International Conference Machine Learning,* 194–202. Morgan Kaufmann, 1995.

[Fayyad and Irani 1993] U.M. Fayyad and K.B. Irani. Multi-interval discretization of continuous-valued attributes for classification learning. *Proceedings of the 13th International Joint Conference Artificial Intelligence (IJCAI '93),* 1022–1027. Chamberry, France, Aug./ Sep. 1993.

[Freitas 1997] A.A. Freitas. The principle of transformation between efficiency and effectiveness: towards a fair evaluation of the cost-effectiveness of KDD techniques. *Proceedings of the 1st European Symposium on Principles of Data Mining and Knowledge Discovery (PKDD '97). Lecture Notes in Artificial Intelligence 1263,* 299–306. Springer, 1997.

[Freitas and Lavington 1996] A.A. Freitas and S.H. Lavington. Speeding up knowledge discovery in large relational databases by means of a new discretization algorithm. *Advances in Databases: Proceedings of the 14th British National Conference on Databases (BNCOD '14). Lecture Notes in Computer Science 1094*, 124–133. Springer, 1996.

[Freitas and Lavington 1998] A.A. Freitas and S.H. Lavington. *Mining Very Large Databases with Parallel Processing*. Kluwer, 1998.

[Ho and Scott 1997] K.M. Ho and P.D. Scott. Zeta: a global method for discretization of continuous variables. *Proceedings of the 3rd International Conference on Knowledge Discovery and Data Mining (KDD '97)*, 191–194. AAAI Press, 1997.

[Hu 1998] Y.-J. Hu. Constructive induction: covering attribute spectrum. In: H. Liu and H. Motoda (Eds.) *Feature Extraction, Construction and Selection: a data mining perspective*, 257–272. Kluwer, 1998.

[Hussain et al. 1999] F. Hussain, H. Liu, C.L. Tan and M. Dash. Discretization: an enabling technique. *Technical Report TRC6/99*. The National University of Singapore. June 1999.

[John et al. 1994] G.H. John, R. Kohavi and K. Pfleger. Irrelevant features and the subset selection problem. *Proceedings of the 11th International Conference Machine Learning*, 121–129. Morgan Kaufmann, 1994.

[Kerber 1992] R. Kerber. ChiMerge: Discretization of numeric attributes. *Proceedings of the 1992 National Conference American Assoc. for Artificial Intelligence (AAAI '92)*, 123–128.

[Kohavi and John 1998] R. Kohavi and G.H. John. The wrapper approach. In: H. Liu and H. Motoda (Eds.) *Feature Extraction, Construction and Selection: a data mining perspective*, 33–50. Kluwer, 1998.

[Kohavi and Sahami 1996] R. Kohavi and M. Sahami. Error-based and entropy-based discretization of continuous features. *Proceedings of the 2nd International Conference Knowledge Discovery and Data Mining (KDD '96)*, 114–119. AAAI, 1996.

[Koller and Sahami 1996] D. Koller and M. Sahami. Toward optimal feature selection. *Proceedings of the 13th International Conference Machine Learning*. Morgan Kaufmann, 1996.

[Liu and Motoda 1998] H. Liu and H. Motoda. *Feature Selection for Knowledge Discovery and Data Mining*. Kluwer, 1998.

[Mannino et al. 1988] M.V. Mannino, P. Chu and T. Sager. Statistical profile estimation in database systems. *ACM Computing Surveys*, 20(3), 191–221, Sep. 1988.

[Paliouras and Bree 1995] G. Paliouras and D.S. Bree. The effect of numeric features on the scalability of inductive learning programs. *Proceedings of the 8th European Conference Machine Learning (ECML '95). LNAI 912*, 218–231. Springer, 1995.

[Parthasarathy et al. 1998] S. Parthasarathy, R. Subramonian and R. Venkata. Generalized discretization for summarization and classification. *Proceedings of the 2nd International Conference on the Practical Applications of Knowledge Discovery and Data Mining (PADD '98)*, 219–239. The Practical Application Company, UK, 1998.

[Pfahringer 1995] B. Pfahringer. Supervised and unsupervised discretization of continuous features. *Proceedings of the 12th International Conference Machine Learning*, 456–463. Morgan Kaufmann, 1995.

[Rendell and Seshu 1990] L. Rendell and R. Seshu. Learning hard concepts through constructive induction: framework and rationale. *Computational Intelligence 6*, 247–270, 1990.

[Richeldi and Rossotto 1995] M. Richeldi and M. Rossotto. Class-driven statistical discretization of continuous attributes. *Proceedings of the 8th European Conference Machine Learning (ECML '95). LNAI 912*, 335–338. Springer, 1995.

[Smyth and Goodman 1991] P. Smyth and R.M. Goodman. Rule induction using information theory. In G. Piatetsky-Shapiro and W.J. Frawley (Eds.) *Knowledge Discovery in Databases*, 159–176. Menlo Park, CA: AAAI Press, 1991.

[Wang and Sundaresh 1998] Wang and Sundaresh. Selecting features by vertical compactness of data. In: H. Liu and H. Motoda (Eds.) *Feature Extraction, Construction and Selection*, 71–84. Kluwer, 1998.

5 Basic Concepts
of Evolutionary Algorithms

"Darwin insisted that his theory explained
not just the complexity of an animal's body
but the complexity of its mind."
[Pinker 1997, p. 22]

This chapter discusses some basic concepts and principles of Evolutionary Algorithms (EAs), focusing mainly on Genetic Algorithms (GAs) and Genetic Programming (GP). The main goal of this chapter is to help the reader who is not familiar with these kinds of algorithm to better understand the next chapters of this book.

This chapter is organized as follows. Section 5.1 presents an overview of EAs and some key issues in their design. Section 5.2 discusses selection methods. Sections 5.3 and 5.4 discuss GAs and GP, respectively. Finally, section 5.5 discusses niching.

5.1 An Overview of Evolutionary Algorithms (EAs)

As mentioned in section 1.3.2, the paradigm of Evolutionary Algorithms (EAs) consists of stochastic search algorithms that are based on abstractions of the processes of Darwinian evolution. Although there are several different kinds of EA (as will be discussed below), almost all EAs have some basic elements in common [Back 2000a; De Jong 2000; De Jong et al. 2000; Hinterding 2000], as follows:

(a) EAs typically work with a population of individuals (candidate solutions) at a time, rather than with a single candidate solution at a time;

(b) They use a selection method biased by fitness, i.e., a measure of quality of the candidate solution represented by an individual. Hence, the better the fitness of an individual, the more often it is selected and the more some parts of its "genetic material", i.e., parts of its candidate solution, will be passed on to later generations of individuals.

(c) They generate new individuals via a mechanism of inheritance from existing individuals. Descendants of individuals are generated by applying stochastic (probabilistic) operators to the existing individuals of the current generation. Two often-used kinds of operator are crossover (recombination) and mutation. In essence, crossover swaps some genetic material between two or more individuals, whereas mutation changes the value of a small part of the genetic material of an individual to a new random value, simulating the erroneous self-replication of individuals.

One can identify four main (sub)paradigms of evolutionary algorithms (EAs), as follows.

Evolution Strategies (ES) typically use an individual representation consisting of a real-valued vector. Early ES emphasized mutation as the main exploratory search operator, but nowadays both mutation and recombination (crossover) are used. An individual often represents not only real-valued variables of the problem being solved but also parameters controlling the mutation distribution, characterizing a self-adaptation of mutation parameters. The mutation operator usually modifies individuals according to a multivariate normal distribution, where small mutations are more likely than large mutations.

Evolutionary Programming (EP) was originally developed to evolve finite-state machines, but nowadays it is often used to evolve individuals consisting of a real-valued vector. Unlike ES, it does not use crossover, in general. Similarly to ES, it also uses normally-distributed mutations and self-adaptation of mutation parameters.

Genetic Algorithms (GA) emphasize crossover as the main exploratory search operator and consider mutation as a minor operator, typically applied with a very low probability. In early, "classic" GAs individuals were represented by binary strings, but nowadays more elaborate representations, such as real-valued strings, are also used. Overall, the similarity between EP and ES is greater than the similarity between GA and EP or ES [De Jong et al. 2000, p. 43], since GA typically emphasize more the simulation of natural genetic mechanisms, whereas EP and ES emphasize more the behavioral relationship between parents and their offspring.

Genetic Programming (GP) is often described as a variation of GA, rather than a mainstream EA paradigm by itself. In this book we prefer to consider GP as a kind of mainstream EA paradigm, for the following reasons. First, the GP community and corresponding research and applications have considerably grown in the last few years. Second, both GA and GP are the focus of this book, since they seem to be the two kinds of EA most used in data mining. Third, in some sense GP is the paradigm of EA that is most distinct from the other ones. After all, the term genetic programming is normally used to indicate that the individuals (candidate solutions) being evolved are some sort of computer programs, consisting not only of data structures but also of functions (or operations) applied to those data structures. The idea of having individuals explicitly representing both data and functions/operators (rather than just parameters of functions) seems at present relatively little used in the other EA paradigms – even though the finite-state machines evolved by early EP systems can be considered as a sort of computer "program" (or algorithm).

As a first approximation, Algorithm 5.1 shows two basic pseudocodes of EAs. More precisely, Algorithm 5.1(a) shows a pseudocode that is representative of GA and GP, whereas Algorithm 5.1(b) shows a pseudocode that is representative of ES and EP. These two pseudocodes are quite similar. The main difference between them is as follows. In GA and GP the selection of good individuals is usually done before the application of genetic operators to create offspring, so that genetic operators are applied to the selected good individuals. In contrast, ES and EP usually apply genetic operators first to create offspring and then select good individuals [Deb 2000].

create initial population;	create initial population;
compute fitness of individuals;	compute fitness of individuals;
REPEAT	REPEAT
select individuals based on fitness;	apply genetic operators to
apply genetic operators to selected	individuals, creating offspring;
individuals, creating offspring;	compute fitness of offspring;
compute fitness of offspring;	select individuals based on fitness;
update the current population;	update the current population;
UNTIL (stopping criteria)	UNTIL (stopping criteria)
(a) Pseudocode for GA and GP	(b) Pseudocode for EP and ES

Algorithm 5.1: Basic pseudocodes for evolutionary algorithms

The above distinctions between EA paradigms should be considered only as general comments, rather than precise statements. Within each of these four paradigms there are many different algorithms, with different characteristics. In addition, there is nowadays a tendency towards the unification of the field of EA. More and more researchers and practitioners of the above four paradigms meet each other in large EA conferences. More and more methods originally developed by the community of one paradigm are adopted by another community. As a result, the distinction between different EA paradigms is blurring.

5.1.1 Key Issues in the Design of EAs

One of the key issues in the design of an EA is the fitness function, since this is the function that will be optimized by the EA. Ideally, it should measure the quality of an individual (candidate solution) as precisely as possible, subject to restrictions of available processing power, knowledge about the problem being solved and the user's requirements. In data mining applications there are several possible rule-quality criteria (section 1.1.1), and it is important to include in the fitness function criteria that are relevant for the user. In addition, if the data set being mined is very large, the fitness of individuals can be computed from data subsets to save processing time (section 11.1). We emphasize that it is important that the values of the fitness function be graded in a fine-grained manner, otherwise the EA will have little information about individual quality to guide its search. For instance, in one extreme a bad fitness function will take on only the value 1 or 0, corresponding to "good" or "bad" individuals, and the EA will be unable to effectively select among many different good individuals.

Another key issue in the design of an EA is to choose an individual representation and genetic operators that are suitable for the target problem. After all, any individual representation and any search operator have a bias, and the effectiveness of any bias is strongly problem-dependent (section 2.5). In addition, the individual representation and the genetic operators must be chosen together, in an synergistic manner, rather than being separately chosen. To quote [Mitchell 1996, p. 174]:

"...it is not a choice between crossover or mutation but rather the balance among crossover, mutation and selection that is all important. The correct balance also depends on details of the fitness function and the encoding. Furthermore, crossover and mutation vary in relative usefulness over the course of a run."

As pointed out by [Schnier and Yao 2000], unfortunately research on individual representations has been less active than research on genetic operators and selection methods. It is hoped that this unbalanced situation will be corrected by the community in the future, by putting more emphasis on investigating the effectiveness of different combinations of individual representations and genetic operators.

Yet another key issue in the design of an EA is to choose an appropriate trade-off between exploitation of the best candidate solutions and exploration of the search space [Eshelman et al. 1989; Whitley 1989; DeJong 2000]. In essence, selection based on fitness is the source of exploitation of the best current candidate solutions, whereas genetic operators such as crossover and mutation are the source of exploration of the search space, creating new candidate solutions.

There are several kinds of selection methods (section 5.2). Different methods have different selective pressures. In essence, selective pressure is related to the takeover time of the selection method, which can be defined as [Deb 2000, p. 170]: *"...the speed at which the best solution in the initial population would occupy the complete population by repeated application of the selection operator alone"*. Hence, a small takeover time implies a strong selective pressure, since the population would soon converge to a highly fit individual. Such individual is quite possibly a locally-optimal (rather than globally-optimal) solution, characterizing a premature convergence. Conversely, a large takeover time implies a weak selective pressure, since the population will take longer to converge to a highly fit individual, making the search more robust and less likely to get trapped in a local optimum.

In general, if the selective pressure is strong it makes sense to use a higher mutation rate, in order to reduce the strong effect of exploitation associated with selection; whereas if the selective pressure is weak it makes sense to use a low mutation rate.

Similarly, there is an interaction between the exploitation associated with selection and the exploratory power of crossover. The higher the selection pressure, the smaller the genetic diversity of a population, and so the smaller the exploratory power of crossover, since crossover can only recombine genetic material that is currently present in the population (subsection 5.3.2).

5.2 Selection Methods

As mentioned above, the essential idea of selection in evolutionary algorithms is that the better the fitness (quality measure) of an individual (candidate solution), the higher must be its probability of being selected, and so the greater the amount of its genetic material that will be passed on to future generations.

In this section we present an overview of three kinds of selection method, namely proportionate selection, ranking and tournament. In proportionate selection the above-mentioned essential idea is usually implemented by a computational simulation of a biased roulette-wheel [Goldberg 1989]. Think of each individual in the population as a slot of a roulette-wheel. This roulette-wheel is biased in the sense that the size of a slot is proportional to the fitness of the corresponding individual. In order to select an individual one simulates a spinning of the roulette-wheel, so that the probability of an individual being selected is proportional to its fitness. In effect, the probability of an individual being selected is given by the ratio of its fitness value (its roulette-wheel slot size) over the total sum of fitness values of all population individuals (the entire size of the roulette-wheel).

This method is simple and intuitive, but it has several disadvantages [Deb 2000]. Firstly, it assumes that all individuals have non-negative (positive or zero) fitness values and that the fitness function must be maximized, rather than minimized. If the fitness function defined by the user does not satisfy any of these two assumptions, that fitness function must be modified to satisfy them. Secondly, it requires the computation of a global statistic (the total sum of fitness values of all population individuals), which reduces the EA's potential for parallel processing. (Methods for parallelizing an EA will be discussed in chapter 11.) Thirdly, if one individual has a fitness value much better than the others, that individual will tend to be selected much more often than the others. If the superperformer individual represents a local optimum (rather than a global optimum) in the solution space, there will be a premature convergence for that suboptimal solution. Fourthly, if most individuals have about the same fitness value (which often occurs in later generations, when the EA is converging), those individuals will have about the same probability of being selected, so that the selection will be almost random. These latter two disadvantages can be mitigated by using some kind of scaling scheme – see, e.g., [Goldberg 1989].

Ranking selection consists of two steps [Whitley 1989; Grefenstette 2000]. First all individuals of the population are ranked according to their fitness values. The ranking is done in either ascending or descending order of fitness values, depending on whether the fitness function is to be minimized or maximized, respectively. Then a selection procedure conceptually similar to proportionate selection is applied based on the rankings, rather than on the original fitness values. The better the ranking of an individual, the higher its probability of being selected. We emphasize that in this method individuals are selected based only on their performance relative to other individuals. Information about the magnitude of fitness differences between individuals is intentionally discarded.

Tournament selection consists of randomly choosing k individuals from the population – where k is a parameter called the tournament size – and let them "play a tournament" [Blickle 2000]. In general only one winner among the k participants is selected. The winner of the tournament can be chosen in a deterministic or probabilistic manner. In the deterministic case the winner is the individual with the best fitness among the k participants. In the probabilistic case each of the k individuals can be chosen as the winner with a probability proportional to its fitness – this is like using a "mini roulette-wheel" referring only to k individuals. Deterministic tournament selection seems more common in practice.

Both ranking selection and tournament selection avoid some of the above-mentioned disadvantages of roulette-wheel selection. They naturally cope with negative fitness values, as well as both minimization and maximization problems. The reason is that in both ranking and tournament selection the original fitness values need only to be compared with each other to determine their relative ordering. In addition, depending on the choice of parameters used in ranking selection and tournament selection, these two selection methods are approximately equivalent [Julstrom 1999].

However, ranking selection still has a disadvantage similar to the above-mentioned second disadvantage of roulette-wheel selection. It requires a global ranking of all individuals, which reduces the EA's potential for parallel processing. Tournament selection does not have this disadvantage, as long as the tournament size k is much smaller than the population size – which is usually the case.

Note that the parameter k of tournament selection can be used to directly control the selective pressure of the method. The larger the value of k, the stronger the selective pressure. In addition, all other things being equal, deterministic tournament has a stronger selection pressure than probabilistic tournament. Actually, this statement can be generalized, i.e., a deterministic selection method normally has a stronger selection pressure than its probabilistic counterpart.

A more detailed discussion about the above-discussed and other kinds of selection methods can be found in [Back et al. 2000, 166-204].

5.3 Genetic Algorithms (GA)

This section is divided into three parts. Subsection 5.3.1 discusses individual representation. Subsections 5.3.2 and 5.3.3 discuss crossover operators and mutation operators, respectively. A more comprehensive, detailed introduction to GA can be found in [Goldberg 1989; Mitchell 1996; Michalewicz 1996].

5.3.1 Individual ("Chromosome") Representation

An individual (or "chromosome") is a candidate solution for the target problem. In GA, an individual is often represented by a fixed-length string of genes, where each gene is a small part of a candidate solution.

In the simplest case an individual is represented by a string of bits, so that each gene can take on either 0 or 1. One problem with the binary alphabet concerns the case where the variables encoded into an individual are continuous (real-valued) variables. In this case the use of a binary alphabet implies a certain loss of precision in the representation of continuous variables. As put by [Back 2000b, p. 133]:

"...not all points of the search space are represented by binary vectors, such that the genetic algorithm performs a grid search and, depending on the granularity of the grid, might fail to locate an optimum precisely."

This loss of precision can be reduced by using a larger number of bits to represent a continuous variable, but the problem is that the length of an individual would increase fast with the number of continuous attributes, and a very long individual length slows down the GA. The direct use of real values into the individual representation avoids this problem.

In addition, in many real-world applications the need for individual representations that are closer to the problem being solved, involving problem-specific data structures, is now well-established. Such problem-specific individual representations are often associated with problem-specific "genetic" operators. As pointed out by [Michalewicz 1996, p. 7]:

"It seems that [classical] GAs ... are too domain independent to be useful in many applications. So it is not surprising that evolution programs, incorporating problem-specific knowledge in the chromosomes' data structures and specific 'genetic' operators, perform much better."

A practical advice for choosing an individual representation is the principle of minimal alphabets [Goldberg 1989, p. 80]:

"The user should select the smallest alphabet that permits a natural expression of the problem."

A related advice is given by [Back 2000b, pp. 134-135]:

"...one might propose the requirement that, if a mapping between representation space and search space is used at all, it should be kept as simple as possible and obey some structure preserving conditions that still need to be formulated as a guideline for finding a suitable encoding."

Finally, it is important to recall two crucial points. First, any individual representation has its own representation bias, and – as discussed in section 2.5 – the effectiveness of any bias is strongly dependent on the problem being solved. Second, the effectiveness of any individual representation is also strongly dependent on the genetic operators to be applied to that representation, as mentioned in subsection 5.1.1.

5.3.2 Crossover Operators

The crossover (or recombination) operator essentially swaps genetic material between two individuals, called parent individuals. Figure 5.1 illustrates a simple form of crossover, called one-point crossover, between two individuals, each represented as a string with six genes. Figure 5.1(a) shows the parent individuals before crossover. A crossover point is randomly chosen, represented in the figure by the symbol "|" between the second and third genes. Then the genes to the right of the crossover point are swapped between the two individuals, yielding the new child individuals (or offspring) shown in Figure 5.1(b).

If each individual represents, say, a candidate prediction rule, and each gene represents a rule condition, crossover would be swapping rule conditions between two prediction rules.

$$X_1\ X_2\ \big|\ X_3\ X_4\ X_5\ X_6 \qquad\qquad X_1\ X_2\ \big|\ Y_3\ Y_4\ Y_5\ Y_6$$
$$Y_1\ Y_2\ \big|\ Y_3\ Y_4\ Y_5\ Y_6 \qquad\qquad Y_1\ Y_2\ \big|\ X_3\ X_4\ X_5\ X_6$$

(a) Before crossover   (b) After crossover

Figure 5.1: Simple example of one-point crossover

$$X_1\ \boxed{X_2}\ X_3\ \boxed{X_4}\ \boxed{X_5}\ X_6 \qquad X_1\ \boxed{Y_2}\ X_3\ \boxed{Y_4}\ \boxed{Y_5}\ X_6$$
$$Y_1\ \boxed{Y_2}\ Y_3\ \boxed{Y_4}\ \boxed{Y_5}\ Y_6 \qquad Y_1\ \boxed{X_2}\ Y_3\ \boxed{X_4}\ \boxed{X_5}\ Y_6$$

(a) Before crossover   (b) After crossover

Figure 5.2: Simple example of uniform crossover

Another form of crossover is uniform crossover [Syswerda 1989; Falkenauer 1999]. In this kind of crossover, for each gene position (or locus) the genes from the two parents are swapped with a fixed, position-independent probability p. This procedure is illustrated in Figure 5.2, where the three swapped genes are denoted by a box. Note that the swapped genes do not need to be adjacent to each other, unlike one-point crossover.

The higher the value of p, the higher the number of genes swapped between the two parents. In the literature the value of p is often set to 0.5, and sometimes this value is not even made explicit – in this case it is just mentioned that genes are randomly swapped. However, it should be noted that the optimal value of p, like virtually any other parameter of a GA (or of a data mining algorithm, for that matter) is strongly dependent on both the problem being solved and the values of the other parameters of the algorithm.

In any case it is important to understand the influence of the value of p in the exploratory power of uniform crossover. The closer p is to 0.5, the larger the number of genes swapped between the two parents, and so the greater the exploratory power of uniform crossover, i.e., the more global the search performed by this operator is. Conversely, the closer p is to 0 or 1, the smaller the number of genes swapped between the two parents, and so the more local the search performed by this operator is.

In addition, the exploratory power of uniform crossover (and other kinds of crossover) also depends on the genetic variety of the population. Let us elaborate on this point. Assume for now that two parents have very different genotypes, which will usually be the case in early generations of the GA, where the population has not converged yet to a single or a few individuals. In this case, suppose that $p = 0.1$ or $p = 0.9$. Then on average each child will have 90% of the genes of one parent and 10% of the genes of the other parent, i.e., each child will be very close, in the genotype space, to one of the parents. Conversely, if $p = 0.5$ each child will be quite distant, in the genotype space, to both parents. Now suppose that two parents have very similar genotypes, which will usually be the case in later generations of the GA, where the population has probably converged to a single or a few individuals. In this case, even with $p = 0.5$, each child will be very close, in the genotype space, to both parents.

It should be noted that in general GA crossover is a "blind" operator, in the sense that the genes to be swapped are randomly chosen without any attempt to maximize the quality (fitness) of the offspring. However, several authors have recently proposed crossover operators that use some kind of heuristic – rather than just random recombination – to try to produce highly-fit offspring. Some examples can be found in [Mohan 1999; Rasheed 1999; Ho et al. 1999; Chen and Smith 1999]. In general, one disadvantage of these "intelligent" kinds of crossover is that they tend to be considerably more computationally expensive than conventional, blind crossover operators.

5.3.2.1 Biases of Crossover Operators

Any genetic operator has a search bias (section 2.5), whose effectiveness is strongly dependent on the application domain. Hence, it is important to understand the biases of genetic operators.

A comprehensive discussion about the biases of several kinds of crossover operators in genetic algorithms can be found in [Eshelman et al. 1989; Booker 1991; Eshelman et al. 1997; Rana 1999; Vekaria and Clack 1999].

Here we focus on two major kinds of bias associated with a crossover operator, namely positional bias and distributional bias. Our discussion here is mainly based on [Syswerda 1989], but all the above references are useful for a better understanding of crossover biases.

A crossover operator has positional bias to the extent that the creation of new individuals by swapping genes of the parents depends on the position of the genes in the genome. A traditional one-point crossover has a strong positional bias. To see why, let L denote the length (number of genes) of the genome encoding an individual, so that a crossover point is randomly chosen, with uniform probability distribution, in the range $1...L-1$. The probability of two adjacent genes, say the third and fourth genes, being swapped together is much higher than the probability of two "distant" genes, say the first and the $(L-1)$-th gene, being swapped together. On the other hand, a uniform crossover has no positional bias. The probability of a gene being swapped is independent of the position of that gene in the genome. In between the two extremes of one-point and uniform crossover, there are multi-point crossover operators, whose positional biases generally decrease as we increase the number of crossover points.

Intuitively, the strong positional bias of one-point crossover does not seem desirable in many applications of GA for prediction-rule discovery. This application of GAs will be discussed in detail in chapter 6. For now just consider the general scenario where each individual represents a candidate IF-THEN prediction rule and each gene represents a condition (say an attribute-value pair) of the rule antecedent (IF part). Assume a positional individual representation where each gene is associated with a given attribute, and the value of that gene specifies the value of the corresponding attribute in the rule antecedent. As mentioned in section 1.2.1, the rule antecedent is usually composed of a logical conjunction of conditions. From a logical viewpoint, these conditions have no ordering, i.e., a rule antecedent contains a (unordered) set of conditions. Of course, when rule conditions are encoded into an individual they have a left-to-right ordering in the genotype, but such ordering is defined only for the purpose of implementing the GA. The position of attributes in the genotype is entirely arbitrary from a phenotypic viewpoint, and it should be ignored when the genotype is decoded into a candidate rule. However, due to its strong positional bias, one-point crossover would tend to disrupt combinations of attributes that are very separated in the genotype and to preserve combinations of attributes that are adjacent in the genotype. This bias can be avoided by using, e.g., uniform crossover, where attribute-value pairs (genes) are swapped or not in a way independent of their position in the genotype.

Of course, if one knows that two or more attributes interact in an important way, one can exploit this knowledge by encoding those attributes in adjacent positions in the genotype and using one-point crossover, with its strong positional bias. However, this requires some previous knowledge about the application domain, which is not available in many real-world data mining applications, where we are essentially searching for unknown relationships in the data.

A crossover operator has distributional bias to the extent that the number of genes that is expected to be swapped between the two parents is distributed around some value or values, rather than being uniformly distributed in the range $1...L\text{-}1$ genes. Traditional one-point crossover has no distributional bias. To see why, note that all the possible numbers of genes to be swapped, in the range $0...L\text{-}1$, are equally likely, since the crossover point is randomly chosen in the range $0...L\text{-}1$. On the other hand, uniform crossover has a strong distributional bias. The expected number of swapped genes is $p.L$, where p is the probability that the genes at any given position in the two parents will be swapped.

5.3.3 Mutation Operators

Mutation is an operator that acts on a single individual at a time. Unlike crossover, which recombines genetic material between two (or more) parents, mutation replaces the value of a gene with a randomly-generated value. Similarly to crossover, mutation is usually a "blind" operator, in the sense that the gene to be mutated and its new value are randomly chosen without any attempt to maximize the fitness of the new individual.

Unlike conventional crossover, mutation can introduce into an individual a gene value that is not even present in the current population. Hence, mutation helps to increase the genetic diversity of the population. In GAs mutation is usually applied with a small probability, typically much smaller than the crossover probability.

In the simplest individual representation, where an individual consists of a fixed-length, binary string, mutation simply inverts the value of a gene (a bit), i.e., it replaces a "0" with a "1" or vice-versa.

When the individual representation is more complex, a mutation operator is often designed specifically for that representation [Michalewicz 1996]. As usual, each operator has its own bias, which is suitable for some problems and unsuitable for others. For instance, consider the case where an individual is a fixed-length string of real-valued numbers. Creep mutation [Davis 1991, p. 66] adds/subtracts a bounded small random amount to/from a gene value with a fixed probability. This operator will be effective when an individual represents a candidate solution that is very close to an optimum in the fitness landscape, since it creates new individuals that are very similar (close) to the original individual. However, as pointed out by [Fogel 2000, p. 239], this operator will not be effective when the individual gets trapped into a locally-optimal region of the fitness landscape that is wider than the operator's bounded amount.

5.4 Genetic Programming (GP)

As mentioned in section 5.1, GP is considered by some authors as a variation of GA. However, there are some important differences between these two (sub)paradigms of evolutionary algorithms.

The main differences concern the issues of individual representation and corresponding genetic operators. Most GP algorithms use a tree-based individual representation. In this case an individual (candidate solution) is represented by a tree, where in general the internal nodes are functions and the terminal nodes are variables of the problem being solved or constants. There are GP algorithms that use other kinds of individual representation [Banzhaf et al. 1998], but our discussion here is limited to tree-based representations, which are by far the most used ones, even though no representation is universally best for all application domains.

Typically, an individual's tree can grow in size and shape in a very dynamical way, unlike the usually fixed-length string of "conventional" GAs. Arguably, however, this is not the main difference between the individual representations of GP and GA. There are many GAs which also use a variable-length individual representation, and in practice many GP systems use some limit of tree depth. Once a tree has a maximum depth, one can view the genome of a tree-based evolutionary algorithm as a full binary tree, expanded out to that depth [Wineberg and Oppacher 1996]. In this case the size of the genome is fixed and conventional GA operators can be applied, but the effective, decoded program can be of variable length if one allows some nodes of the full binary tree to be non-executed code.

Arguably, a more important distinction is that in GP an individual can contain not only values of variables (data) but also functions; unlike a GA individual, which typically contains only data. Hence, an individual can be called a "program", though sometimes in a loose sense of the word. In particular, in principle an individual should correspond to a general "recipe" for solving a given kind of problem, whereas a GA individual typically corresponds to a solution for one particular problem instance.

In any case, as mentioned in section 5.1, the distinction between GA and GP (and other kinds of EAs) is blurring as the tendency towards the unification of the field of EAs increases.

Concerning GP terminology, it is worthwhile to mention that [Montana 1995] suggests using the term "nonterminal set" to describe what [Koza 1992] calls the "function set", since the terminal set can include a function (or subroutine) that returns a value, as long as the function takes no argument. For instance, the terminal set of a GP often includes a random-constant generator. Although the term "nonterminal set" is probably better from the viewpoint of technical correctness, the term "function set" is much more used in practice, due to the influence of Koza's books. Hence, in this book we will use the term "function set" to be consistent with the majority of the GP literature.

The rest of this section is divided into five subsections. Subsection 5.4.1 discusses the function set of GP. Subsections 5.4.2 and 5.4.3 discuss crossover operators and mutation operators, respectively. Subsection 5.4.4 discusses the standard GP approach for classification. (More elaborate GP approaches for classification, which seem more suitable in the context of data mining, will be discussed in chapter 7.) Finally, subsection 5.4.5 discusses the problem of code bloat. A more comprehensive introduction to GP can be found in [Koza 1992] and [Banzhaf et al. 1998].

5.4.1 Function Set

In general the function set of a GP algorithm must satisfy at least two properties, namely sufficiency and closure [Koza 1992]. Sufficiency means that the function set's expressive power is good enough to be able to represent a solution to the target problem.

Closure means that a function should be able to accept, as input, any output produced by any function in the function set. In practice there are several situations where GP has to cope with variables having different data types, which makes it difficult to satisfy the closure property. Good solutions for this problem are not trivial.

Consider, for instance, an approach where a function signals an error when its arguments are of inconsistent type, and then an infinitely bad evaluation is assigned to the corresponding tree. The problem is that this approach can be very inefficient, spending most of its time evaluating syntactically-illegal trees. For instance, [Montana 1995, p. 221] reports that, in a multidimensional least squares regression problem involving matrices and vectors, in an initial population of 50,000 individuals only on the order of 20 individuals were type consistent.

Another approach is to have functions cast a data type into another so that the execution of a function becomes valid. It might be argued that this approach works reasonably well when there is a natural way to cast one data type into another. However, this approach can lead to bad results when there is no natural way to cast the actual data type of an argument of a function into the data type required by that function. [Montana 1995] points out that even if unnatural data type conversions succeed in finding a solution for a particular set of data, they are unlikely to be part of a symbolic expression that can generalize to new data.

In the context of data mining, it should be noted that in general cast operations should not be allowed in database systems. In principle in these systems each attribute is (or should be) associated with a *domain*, which is essentially a *data type* definition that restricts the kinds of operations that can be applied to values of the attribute [Date 2000, chap. 5]. For instance, suppose a database containing records about customers includes the integer-valued attribute *Age* and the real-valued attribute *Salary*. From a syntactical viewpoint we can naturally cast an integer number into a real number to perform a given arithmetic operation, such as *Age + Salary*. However, this kind of arithmetic comparison obviously does not make any sense at all, from a semantical viewpoint. That is what database domains are good for. They are a semantic aspect of data integrity in a database system.

The issue of how to satisfy the closure property in the context of data mining requires a detailed discussion, which is postponed to chapter 7.

In addition to the above two properties of a function set, one could add a third one. Ideally a function set should be parsimonious, in the sense that it should contain only the functions that are necessary to solve the target problem. In theory this third property is not so crucial, since GP should be able to automatically generate solutions containing only relevant functions for the target problem. In practice, however, a parsimonious function set is important because the size of the search space usually increases exponentially with the number of functions. Hence, the more irrelevant functions we include in the function set, the less cost-effective the GP search will be.

The problem, of course, is that in many interesting real-world applications we rarely know exactly which are the necessary functions. Actually, if we already knew precisely which functions are necessary for the target problem, it is quite possible that there is a special-purpose problem-solving technique that could take advantage of this knowledge to solve the problem in a more cost-effective way than GP.

Hence, in practice the choice of functions to be included in the function set usually involves a trade-off between the expressive power of solutions and the cost-effectiveness of the search. On one hand, the larger the number of functions included in the function set, the higher the expressive power available to generate good solutions, but the larger the search space, and so the harder is to find a good solution in the middle of so many candidate solutions. On the other hand, the smaller the number of functions included in the function set, the smaller the search space, but the smaller the expressive power available to generate good solutions.

To cope with this trade-off, [Banzhaf et al. 1998, p. 111] suggest the following:

"An approximate starting point for a function set might be the arithmetic and logic operations: PLUS, MINUS, TIMES, DIVIDE, OR, AND, XOR. The range of problems that can be solved with these functions is astonishing. Good solutions using only this function set have been obtained on several different classification problems, robotics control problems, and symbolic regression problems."

5.4.2 Crossover Operators

Like GA crossover, GP crossover essentially swaps genetic material between two parent individuals. However, since the individual representation used by GP is typically a tree or another variable-length representation, GP crossover has to adapted to the used individual representation. Tree crossover usually replaces a randomly-selected subtree of an individual with a randomly-chosen subtree from another individual. This procedure is illustrated in Figure 5.3, where the "crossover point" is indicated by a tilted line and the subtrees swapped by crossover are shown in bold.

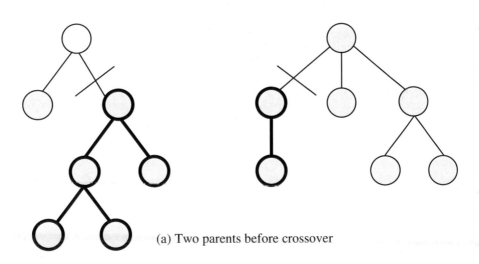

(a) Two parents before crossover

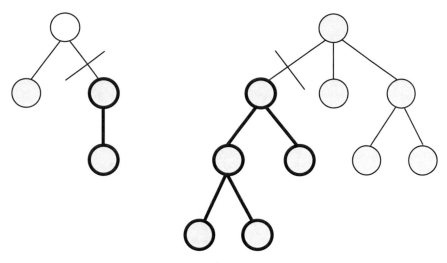

(b) Two children (offspring) produced by crossover

Figure 5.3: Conventional tree crossover in genetic programming

Note that this kind of crossover, hereafter called conventional tree crossover, is essentially a "blind", context-insensitive operator. The term "blind" operator is here used to denote the fact that crossover points are randomly chosen without any attempt to maximize the quality of the offspring, similarly to GA crossover.

The term context-insensitive denotes the fact that, when conventional tree crossover moves a subtree S from one individual i_1 to another individual i_2, it ignores the context in which S will be put. Suppose that S contributes for ameliorating the fitness of i_1, i.e., S is a good "building block", in the sense of being a highly-fit, small piece of genetic material. This quality of S is probably dependent on its context, i.e., on the other tree nodes in i_1. When S is moved to i_2 it will be probably put into a very different context. In this new context it is quite possible that S contributes for worsening the fitness of i_2.

The problem that the quality of small pieces of genetic material is context-dependent can also occur in GAs. However, it seems that the problem is more serious in GP. The difference in the magnitude of the problem in these two cases can be justified as follows.

GAs usually work with fixed-length, positional-semantics genomes. The term positional semantics denotes the fact that the semantics (or meaning) of a gene is determined by its position (or locus, in GA terminology) in the genome. As a result of this positional semantics, most GA crossover operators at least swap homologous genes between two individuals.

In contrast, GP usually works with variable-length genomes, without positional semantics. This lack of positional semantics greatly reduces conventional tree crossover's ability in combining good building blocks of parent individuals, as pointed out by [Angeline 1997, p. 16]:

"The lack of positional semantics is one of the distinguishing features of genetic programs ... and is the most compelling reason to discount [conventional] subtree crossover's ability to be a building block engine."

The above problems suggest that conventional tree crossover, as well as other GP crossovers that are blind and context-insensitive, usually tend to have a certain destructive effect on the fitness of the spring, by comparison with the fitness of the parents. Quoting [Banzhaf et al. 1998, p. 153]:

"The effect of crossover has been measured for tree-based GP, linear (machine code) GP ..., and graph GP ... . In all three cases, crossover has an overwhelmingly negative effect on the fitness of the offspring of the crossover. .... crossover routinely reduces the fitness of offspring substantially relative to their parents in almost every GP system. This stands in stark contrast to biological crossover."

There has also been some studies explicitly designed to evaluate whether or not conventional tree crossover is a kind of macromutation operator restricted by population content. The term "macromutation" not only emphasizes that GP crossover produces offspring which are syntactically very different from the parents but also suggests the idea of destructiveness inherent in a mutation operator. The term "restricted by population content" reminds us that, unlike a true mutation operator, where the genetic material to be inserted into the offspring is randomly generated, crossover only inserts into the offspring genetic material that is already present in the population.

Such studies usually compare the performance of crossover with the performance of a kind of true macromutation operator called headless chicken crossover (HCC) [Banzhaf et al. 1998, pp. 153-155; Angeline 1997]. The basic idea of this operator is that one parent is chosen from the current population, but the other parent is produced by randomly generating an entire new individual. The two parents are then crossed over as usual, producing one offspring. As a result, HCC effectively replaces part of the genetic material of a preexisting individual by a randomly-generated genetic material.

In several experiments reported in the above-mentioned references a HCC operator produced results that were about the same or somewhat better than conventional tree crossover. These studies provide support to the viewpoint that conventional tree crossover is a kind of macromutation operator restricted by population content.

The above problems associated with conventional tree crossover have led to several suggestions for improving this operator. For instance, Banzhaf et al. dedicated 15 pages of their GP textbook [Banzhaf et al. 1998, pp. 156-170] to discussing several proposals for improving crossover.

Another proposal for improving GP crossover was made by [Poli and Langdon 1998; Poli et al. 1999], who describe a GP uniform crossover operator. This operator is inspired by the GA operator of the same name. The basic idea is as follows [Poli et al. 1999, pp. 1163-1164]:

"GP uniform crossover begins with the observation that many parse trees are at least partially structurally similar. This means that if we start at the root node and work our way down each tree, we can frequently go some way before finding function nodes of different arity at the same locations. Furthermore we can swap every node up to this point with its counterpart in the other tree without altering the structure of either."

At a high level of abstraction GP uniform crossover is based on the idea of identifying the parts of the two parent trees that have similar structure and then trying to preserve that structure by choosing suitable crossover points. GP uniform crossover, like other GP context-preserving crossover operators proposed by [D'haeseleer 1994], can be considered an homologous crossover with respect to tree structure, i.e., with respect to the syntax of the tree, but not necessarily with respect to its semantics (behavior).

5.4.3 Mutation Operators

Like GA mutation, GP mutation is an operator that acts on a single individual at a time, and it is often applied with a relatively low probability. Actually, Koza deemed mutation a largely unnecessary operator in GP, and did not use it in the majority of the experiments reported in his first GP book, justifying this position as follows [Koza 1992, pp. 106-107]:

"First, in genetic programming, particular functions and terminals are not associated with fixed positions in a fixed structure. Moreover, when genetic programming is used, there are usually considerably fewer functions and terminals for a given problem than there are positions in the conventional genetic algorithm. Thus, it is relatively rare for a particular function or terminal ever to disappear entirely from a population in genetic programming. Therefore, to the extent that mutation serves the potentially important role of restoring lost diversity in a population for conventional genetic algorithm, it is simply not needed in genetic programming. Second, in genetic programming, whenever the two crossover points in the two parents happen to be endpoints of trees, the crossover operation operates in a manner very similar to point mutation. Thus, to the extent that point mutation may be useful, the crossover operator already provides it."

It should be noted, however, that even with a very large population size, it is possible that a population does not contain a symbol that is very relevant for obtaining the optimal solution, particularly in problems involving real-valued constants, where the number of possible constants is infinite. For instance, Angeline observed that some form of mutation is desirable in a fully-numeric sunspot-prediction problem, as follows [Angeline 1997, p. 15]:

"Finding good solutions for this problem is highly dependent on having appropriate numeric terminals in the population's trees. Without some form of mutation, subtree crossover is doomed to substandard performance on this problem

since it is forced to use only a subset of the numeric constants introduced in the initial population. ... The macromutations are not susceptible to this problem since they have the opportunity to introduce new numeric terminals with each application that may be more beneficial than the current numeric terminals."

As one approach to improve the way GP copes with numeric constants, [Evett and Fernandez 1998] proposed a numeric mutation operator that essentially replaces all numeric constants of an individual with new randomly-chosen constants.

In addition, several experiments have shown that some form of mutation is often as effective as tree crossover in finding good solutions [Angeline 1997; Fuchs 1998; Luke and Spector 1997, 1998].

We briefly review below four types of GP mutation. These and other types of GP mutation are also discussed in [Banzhaf et al. 1998, pp. 240-243; Cavaretta and Chellapilla 1999].

Point mutation - This is the simplest form of mutation. It consists of replacing a single node in a tree with another randomly-generated node of the same kind. By "same kind" we mean that, if the node to be replaced is an internal node, the new randomly-generated node should be able to accept as its children nodes the current children nodes of the node to be replaced. If the node to be replaced is a terminal node, it is simply replaced by another terminal node. Point mutation replacing a terminal node is illustrated in Figure 5.4.

Collapse mutation - This operator randomly selects an internal node in the tree, and then it replaces it with a new randomly-generated terminal node. Collapse mutation is illustrated in Figure 5.5.

Expansion mutation - This operator randomly selects a leaf (terminal) node in the tree, and then it replaces it with a new randomly-generated subtree. Expansion mutation is illustrated in Figure 5.6.

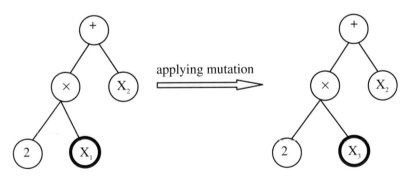

Figure 5.4: Example of point mutation

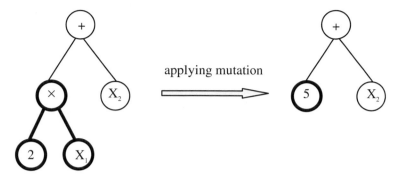

Figure 5.5: Example of collapse mutation

Subtree mutation - This operator randomly selects an internal node in the tree, and then it replaces the subtree rooted at that node with a new randomly-generated subtree. The new randomly-generated subtree should be created subject to some restrictions of depth and/or size, to avoid the generation of a too large subtree. One often-used restriction is that the size and/or depth of this new subtree should be subject to the same limitations as the trees generated in the initial random population. Subtree mutation is illustrated in Figure 5.7.

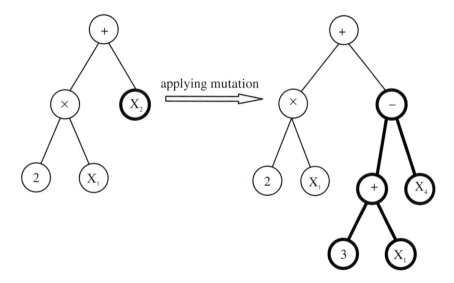

Figure 5.6: Example of expansion mutation

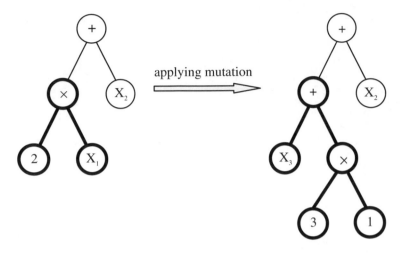

Figure 5.7: Example of subtree mutation

Table 5.1: Influence of different mutations on the size of the tree being mutated

kind of mutation	effect on the size of the tree being mutated
point	unaltered
collapse	reduced
expansion	increased
subtree	unpredictable

Note that each of these mutation operators has a bias. In particular, they have different influences on the size of the tree being mutated, as summarized in Table 5.1. For instance, collapse mutation has the effect of reducing the size of the tree being mutated, i.e., it has a bias favoring smaller trees. Expansion mutation has the opposite effect, increasing the size of the tree being mutated, i.e., it has a bias favoring larger trees. Therefore, the decision to use collapse mutation or expansion mutation should be made by taking into account the interaction between these biases and several other factors related to both the application domain and other GP operators.

For instance, suppose that the fitness function already includes a parsimony pressure (favoring small trees). In this case the use of collapse mutation (which also favors small trees) would be somewhat redundant. In particular, in this case collapse mutation could not only result in too much overall parsimony pressure but also make it more difficult to determine which components of GP should be assigned the credit (blame) for discovering (or not) a simple, accurate solution. In any case we emphasize that the decision about which mutation operator (if any) should be used is strongly dependent on the problem being solved.

5.4.4 The Standard GP Approach for Classification

The standard GP approach for classification is essentially a GP that performs symbolic regression. Like classification, regression can be considered a data mining task involving the prediction of the value of a user-defined goal attribute, given the values of the other (predictor) attributes. The main difference between classification and regression is the following. In classification the goal attribute is categorical (nominal), i.e., it can take on a small number of discrete values, or classes. In contrast, in regression the goal attribute is continuous (real-valued).

In both tasks the set of predictor attributes can include both categorical and continuous attributes. In practice, however, in the regression task it is probably more common to have a set of predictor attributes containing only continuous attributes. This allows the use of a wide range of statistical/numerical regression methods.

Similarly, most applications of GP for symbolic regression assume that all predictor attributes are continuous. In this case one can simply include in the function set several kinds of mathematical function appropriate to the application domain and include in the terminal set the predictor attributes and a random-constant generator [Koza 1992, chap. 10]. To make a prediction for a given data instance, the predictor attributes at the leaves of a GP individual are instantiated with the corresponding values in that data instance. Once we apply the functions in the internal nodes of a GP individual to the values of the attributes in the leaf nodes of that individual, the system computes a numerical value that is output at the root node of the tree. This output is the goal-attribute value predicted by GP for the predictor-attribute values of the data instance which were given as input to the GP.

As long as all the predictor attributes are continuous, this form of GP for symbolic regression can be easily adapted for classification by using a simple "trick". The basic idea is to assign numerical values to the classes to be predicted and then predict the class whose value is the closest to the numeric value output at the root of the tree. This idea is most often used to discriminate between two classes, so that if the value output at the root of the tree is greater than a given threshold the system predicts a given class, otherwise the system predicts the other class. For instance, assuming the value output at the root of the tree is normalized to be in the range [0..1], we can assign to one class the value 0 and to the other class the value 1. Hence, if the value output at the root of the tree is less than 0.5 we predict class 0, otherwise we predict class 1.

Note that this approach ignores how close the value output at the root node is to 0 or to 1. For instance, an output value of either 0.1 or 0.4 would be used to predict class 0. Intuitively, this approach ignores potentially-relevant information, namely the actual magnitude of the value output at the root node. One approach to take into account this magnitude can be found in [Muruzabal et al. 2000].

If the data set has C classes, $C > 2$, the problem is usually transformed into C two-class problems [Kishore et al. 2000]. In this case the GP is run C times. The i-th run, $i=1,...,C$, evolves an expression for discriminating between class c_i and all other classes. Then all C evolved expressions are used for predicting the class of new data instances. Of course, this requires some form of conflict resolution between the *C* evolved expressions when two or more expressions have their conditions satisfied by a given data instance.

Instead of transforming a *C*-class problem into *C* two-class problems, one can directly solve a *C*-class problem with a single run of GP. The basic idea is that different individuals (of the same generation) represent rules predicting different classes. In this case the GP search mechanism would probably have to be somewhat modified to enforce population diversity [Freitas 1997]. For instance, in principle it would be desirable to enforce that there is at least one individual predicting each class in the current population. Some kind of niching method (section 5.5) could be applied at the level of predicted class – each class could be considered a niche, so that GP would be forced to generate individuals of different classes, to cover different niches in data space. This approach of using a single run of GP solving a *C*-class problem would probably be more complex to design and implement than the approach of using *C* runs of GP solving a two-class problem, but the former approach would be considerably faster, particularly as the number of classes (*C*) increases.

In any case, the standard GP approach for classification, including in the function set mainly mathematical functions, does not naturally address the case where the terminal set contains both continuous and categorical attributes. In addition, it does not produce high-level, comprehensible prediction rules, in the form of the rules discussed in section 1.2.1. The use of GP for discovering comprehensible prediction rules will be discussed in chapter 7.

5.4.5 Code Bloat

In many applications, the size of GP trees tend to grow fast without a corresponding improvement in fitness. This growth, often called "code bloat" in the literature, consists mainly of code that does not change the semantics of the evolving individuals. Several causes of code bloat are discussed in [Langdon et al. 1999].

Some examples of pieces of code that do not change the semantics of a GP individual are "X + 0" and "X OR X". Such pieces of code are often called "introns", by analogy with the biological concept of introns [Banzhaf et al. 1998, chap. 7].

It should be emphasized that code bloat can occur even without introns, depending on how one defines code bloat. For example, [Angeline 1998] defines code bloat as follows:

"Bloat ... can be defined as the expression of a behavior using an excessive amount of the provided symbols relative to the minimum number of those symbols required to express the same behavior. For instance, if the desired behavior is to output the value "4" then the structure "1+1+1+1" expresses that behavior with more symbols than the structure "2+2" and is thus more bloated."

As pointed out by Angeline, code bloat (in the above sense) in general does not serve any useful purpose in an EA. In contrast, introns may have some potentially-beneficial effects, such as protecting the effective code of an individual (the code that does change the semantics of an individual) from the destructive effects of GP crossover. The basic idea is that, if the majority of an individual's code consists of introns, there is a relatively small probability that GP crossover will exchange part of the effective code of an individual with another individual.

However, even if introns have this beneficial effect, they still have some undesirable effects related to the problem of code bloat in general, such as making it difficult to improve the effective code of existing individuals and increasing the processing time and memory usage of GP. As a result, several methods for reducing code bloat have been proposed [Banzhaf et al. 1998, chap. 7; Langdon et al. 1999]. One of these methods involve the use of a parsimony pressure in the fitness function, penalizing larger individuals.

For instance, [Soule et al. 1996] report that penalizing larger individuals has considerably reduced code growth and the percentage of non-functional code without unduly reducing the effectiveness of the solutions produced by GP. They also reported that this approach was more successful than explicitly removing non-functional code via an editing operator. Their result was obtained on a single case study, namely a robot guidance problem.

On the other hand, it can be argued that determining exactly how much an individual should be penalized for its large size is an "art" [Rosca 1996]. For instance, [Soule et al. 1996] suggests that penalization of large individuals should be applied only when the individual's size is larger than a given threshold, which is ideally close to the size of the shortest program which could solve the task. Otherwise it is quite possible that individuals will, at least to a certain degree, evolve to have small sizes, rather than to represent a good solution for the target problem. The problem, of course, is that in practice we rarely know which is the size of the shortest individual that could solve the task. In any case, including a parsimony pressure in the fitness function is a simple method to reduce code bloat and favor smaller trees.

5.5 Niching

In multimodal problems, where the fitness landscape contains several peaks (optima), a conventional evolutionary algorithm often converges to a single optimum. There are at least two problems associated with this convergence. First, in the context of optimization, it may be a premature convergence, i.e., the algorithm may converge to a local optimum before it had time to explore another part of the fitness landscape containing the global optimum. A common response to this problem is to modify the algorithm parameters that control the trade-off between exploitation and exploration, in order to avoid the premature convergence problem.

The second problem is somewhat more subtle. At least in some problems, we do want to discover several optima, rather than just "the" (or one of the) global optima. In these cases, the goal is not only to avoid premature convergence, but to avoid convergence throughout evolution. More precisely, we want to avoid "total" convergence, where virtually all individuals converge to a single optimum. What we want is a form of "partial", "fitness-proportionate" convergence, where there is a stable subpopulation clustered about each of the peaks (optima), and the size of each subpopulation is proportional to the fitness value associated with the corresponding peak [Goldberg and Richardson 1987]. This corresponds to individuals being spread across several "niches" in the fitness landscape.

Problems where we want this kind of "partial" convergence, so that individuals cover different parts of the problem space, are not rare in data mining. For instance, in the classification task we usually want to discover a *set* of rules covering as much as possible of the data space. This may require some niching-like modifications to an evolutionary algorithm for rule discovery if an individual represents a single rule rather than a set of rules, as will be discussed in chapter 6.

The need to keep population diversity and foster the discovery of several high-quality peaks in the fitness landscape has led to the development of several niching methods [Mafhoud 1995; Mahfoud 2000]. We briefly review below fitness sharing, which is one of the niching methods most used in practice.

The basic idea of the method involves the notion of a sharing function [Goldberg and Richardson 1987; Deb and Goldberg 1989]. In essence, a sharing function is a way of determining how much the payoff of an individual should be degraded due to a neighbor at some distance, measured according to a given distance metric. This is analogous to what occurs in nature, where a niche has a limited amount of resources, and the more individuals there are in a niche, the less resources are available for each individual in that niche.

More precisely, the definition of sharing function proposed by [Goldberg and Richardson 1987] is:

$$\text{sh}(d_{ij}) = 1 - (d_{ij} / distthrs)^a, \text{ if } d_{ij} < distthrs$$
$$\text{sh}(d_{ij}) = 0 \text{ otherwise}$$

where d_{ij} is a measure of distance between individuals i and j, *distthrs* is a distance threshold, and a is a parameter (often set to 1) that determines the shape of the sharing function. The shared fitness of an individual i, denoted f_i', is then defined as its original (potential) fitness, denoted by f_i, divided by its niche count, denoted by n_i'; i.e.: $f_i' = f_i / n_i'$. Finally, the niche count n_i' for an individual i is computed as the sum of all share function values computed for the entire population, i.e.:

$$n_i' = \sum_{j=1}^{P} \text{sh}(d_{ij}), \qquad \text{where } P \text{ is the number of individuals in the population.}$$

Therefore, the sharing function has the role of reducing the fitness of an individual i according to the number of neighboring individuals and their closeness to i.

The "distance" between individuals can be measured either in genotypic space or in phenotypic space. For instance, in the data mining task of classification, where an individual corresponds to a classification rule or rule set, a measure of phenotypic distance between two individuals can be based on the data instances that are covered by both individuals. If both individuals cover almost the same set of data instances their phenotypic distance would be small, whereas if each of the two individuals covers many data instances that are not covered by the other the phenotypic distance would be large.

Note that fitness sharing is a computationally expensive method, since it requires the computation of distance between each individual and every other individual of the population. (However, note that fitness sharing requires the extra computation of distances only, and not of new fitness evaluations. Hence, as long as the number of individuals is relatively small and computing distances between individuals is much faster than computing fitness values, the extra processing time associated with fitness sharing will not be a major problem.) In any case, if necessary sharing functions can be computed from samples of the population, as suggested by [Goldberg and Richardson 1987, p. 48].

References

[Angeline 1997] P.J. Angeline. Subtree crossover: building block engine or macromutation? *Genetic Programming 1997: Proceedings of the 2nd Annual Genetic Programming Conference*, 9–17. Morgan Kaufmann, 1997.

[Angeline 1998] P.J. Angeline. Subtree crossover causes bloat. *Genetic Programming 1998: Proceedings of the 3rd Annual Genetic Programming Conference,* 745–752. Morgan Kaufmann, 1998.

[Back 2000a] T. Back. Introduction to Evolutionary Algorithms. In: T. Back, D.B. Fogel and T. Michalewicz (Eds.) *Evolutionary Computation 1: Basic Algorithms and Operators*, 59–63. Institute of Physics Publishing, 2000.

[Back 2000b] T. Back. Binary strings. In: T. Back, D.B. Fogel and T. Michalewicz (Eds.) *Evolutionary Computation 1: Basic Algorithms and Operators*, 132–135. Institute of Physics Publishing, 2000.

[Back et al. 2000] T. Back, D.B. Fogel and T. Michalewicz (Eds.) *Evolutionary Computation 1: Basic Algorithms and Operators*. Institute of Physics Publishing, 2000.

[Banzhaf et al. 1998] W. Banzhaf, P. Nordin, R.E. Keller, and F.D. Francone. *Genetic Programming – an Introduction: On the Automatic Evolution of Computer Programs and Its Applications*. Morgan Kaufmann, 1998.

[Blickle 2000] T. Blickle. Tournament selection. In: T. Back, D.B. Fogel and T. Michalewicz (Eds.) *Evolutionary Computation 1: Basic Algorithms and Operators*, 181–186. Institute of Physics Publishing, 2000.

[Booker 1991] L.B. Booker. Recombination distributions for genetic algorithms. In: D. Whitley (Ed.) *Foundations of Genetic Algorithms 2*, 29–44. Morgan Kaufmann, 1993.

[Cavaretta and Chellapilla 1999] M.J. Cavaretta and K. Chellapilla. Data mining using genetic programming: the implications of parsimony on generalization error. *Proceedings of the 1999 Congress on Evolutionary Computation (CEC '99)*, 1330–1337. IEEE Press, 1999.

[Chen and Smith 1999] S. Chen and S.F. Smith. Introducing a new advantage of crossover: commonality-based selection. *Proceedings of the Genetic and Evolutionary Computation Conference (GECCO '99)*, 122–128. Morgan Kaufmann, 1999.

[Date 2000] C.J. Date. *An Introduction to Database Systems,* 7th edn. Addison-Wesley, 2000.

[Davis 1991] L. Davis (Ed.) *Handbook of Genetic Algorithms.* Van Nostrand Reinhold, 1991.

[Deb 2000] K. Deb. Introduction to selection. In: T. Back, D.B. Fogel and T. Michalewicz (Eds.) *Evolutionary Computation 1: Basic Algorithms and Operators*, 166–171. Institute of Physics Publishing, 2000.

[Deb and Goldberg 1989] K. Deb and D.E. Goldberg. An investigation of niche and species formation in genetic function optimization. *Proceedings of the 2nd International Conference Genetic Algorithms (ICGA '89)*, 42–49.

[De Jong 2000] K. De Jong. Evolutionary computation: an unified overview. *2000 Genetic and Evolutionary Computation Conference Tutorial Program*, 471–479. Las Vegas, NV, USA, 2000.

[De Jong et al. 2000] K. De Jong, D.B. Fogel and H.-P. Schwefel. A history of evolutionary computation. In: T. Back, D.B. Fogel and T. Michalewicz (Eds.) *Evolutionary Computation 1: Basic Algorithms and Operators*, 40–58. Institute of Physics Publishing, 2000.

[D'haeseleer 1994] P. D'haeseleer. Context preserving crossover in genetic programming. *Proceedings of the 1994 IEEE World Congress on Computational Intelligence*, 256–261. IEEE Press, 1994.

[Eshelman et al. 1989] L.J. Eshelman; R.A. Caruana and J.D. Schaffer. Biases in the crossover landscape. *Proceedings of the 2nd International Conference Genetic Algorithms (ICGA '89)*, 10–19. 1989.

[Eshelman et al. 1997] L.J. Eshelman, K.E. Mathias and J.D. Schaffer. Crossover operator biases: exploiting the population distribution. *Proceedings of the 7th International Conference on Genetic Algorithms (ICGA '97)*, 354–361. Morgan Kaufmann, 1997.

[Evett and Fernandez 1998] M. Evett and T. Fernandez. Numeric mutation improves the discovery of numeric constants in genetic programming. *Genetic Programming 1998: Proceedings of the 3rd Annual Conference (GP '98)*, 66–71. Morgan Kaufmann, 1998.

[Falkenauer 1999] E. Falkenauer. The worth of the uniform. *Proceedings of the Congress on Evolutionary Computation (CEC '99)*, 776–782. IEEE, 1999.

[Fogel 2000] D.B. Fogel. Real-valued vectors (in Mutation Operators chapter). In: T. Back, D.B. Fogel and T. Michalewicz (Eds.) *Evolutionary Computation 1: Basic Algorithms and Operators*, 239–243. Institute of Physics Publishing, 2000.

[Freitas 1997] A.A. Freitas. A genetic programming framework for two data mining tasks: classification and generalized rule induction. *Genetic Programming 1997: Proceedings of the 2nd Annual Conference (GP '97)*, 96–101. Morgan Kaufmann, 1997.

[Fuchs 1998] M. Fuchs. Crossover versus mutation: an empirical and theoretical case study. *Genetic Programming 1998: Proceedings of the 3rd Annual Conference (GP '98)*, 78–85. Morgan Kaufmann, 1998.

[Goldberg 1989] D.E. Goldberg. *Genetic Algorithms in Search, Optimization and Machine Learning.* Addison-Wesley, 1989.

[Goldberg and Richardson 1987] D.E. Goldberg and J. Richardson. Genetic algorithms with sharing for multimodal function optimization. *Proceedings of the International Conference Genetic Algorithms (ICGA '87)*, 41–49. 1987.

[Grefenstette 2000] J. Grefenstette. Rank-based selection. In: T. Back, D.B. Fogel and T. Michalewicz (Eds.) *Evolutionary Computation 1: Basic Algorithms and Operators*, 187–194. Institute of Physics Publishing, 2000.

[Hinterding 2000] R. Hinterding. Representation, mutation and crossover issues in evolutionary computation. *Proceedings of the 2000 Congress on Evolutionary Computation (CEC '2000)*, 916–923. IEEE, 2000.

[Ho et al. 1999] S.-Y. Ho, L.-S. Shu and H.-M. Chen. Intelligent genetic algorithm with a new intelligent crossover using orthogonal arrays. *Proceedings of the Genetic and Evolutionary Computation Conference (GECCO '99)*, 289–296. Morgan Kaufmann, 1999.

[Julstrom 1999] B.A. Julstrom. It's all the same to me: revisiting rank-based probabilities and tournaments. *Proceedings of the Congress on Evolutionary Computation (CEC '1999)*, 1501–1505. IEEE, 1999.

[Kishore et al. 2000] J.K. Kishore, L.M. Patnaik, V. Mani and V.K. Agrawal. Application of genetic programming for multicategory pattern classification. *IEEE Transactions on Evolutionary Computation 4(3)*, 242–258, Sep. 2000.

[Koza 1992] J.R. Koza. *Genetic Programming: on the programming of computers by means of natural selection.* MIT Press, 1992.

[Langdon et al. 1999] W.B. Langdon, T. Soule, R. Poli and J.A. Foster. The evolution of size and shape. In: L. Spector, W.B. Langdon, U-M O'Reilly and P.J. Angeline (Eds.) *Advances in Genetic Programming 3*, 163–190. MIT Press, 1999.

[Luke and Spector 1997] S. Luke and L. Spector. A comparison of crossover and mutation in genetic programming. *Genetic Programming 1997: Proceedings of the 2nd Annual Conference,* 240–248. Morgan Kaufmann, 1997.

[Luke and Spector 1998] S. Luke and L. Spector. A revised comparison of crossover and mutation in genetic programming. *Genetic Programming 1998: Proceedings of the 3rd Annual Conference,* 208–213. Morgan Kaufmann, 1998.

[Mahfoud 1995] S.W. Mahfoud. Niching methods for genetic algorithms. (Ph.D. thesis) *IlliGAL Report No. 95001.* University of Illinois at Urbana-Champaign, May 1995.

[Mahfoud 2000] S.W. Mahfoud. Niching methods. In: T. Back, D.B. Fogel and T. Michalewicz (Eds.) *Evolutionary Computation 2: Advanced Algorithms and Operators*, 87–92. Institute of Physics Publishing, 2000.

[Michalewicz 1996] Z. Michalewicz. *Genetic Algorithms + Data Structures = Evolution Programs,* 3rd edn. Springer, 1996.

[Mitchell 1996] M. Mitchell. *An Introduction to Genetic Algorithms.* MIT Press, 1996.

[Mohan 1999] C.K. Mohan. Crossover operators that improve offspring fitness. *Proceedings of the Congress on Evolutionary Computation (CEC '99)*, 1542–1549. IEEE, 1999.

[Montana 1995] D.J. Montana. Strongly Typed Genetic Programming. *Evolutionary Computation 3(2)*, 199–230, 1995.

[Muruzabal et al. 2000] J. Muruzabal, C. Cotta-Porras and A. Fernandez. Some probabilistic modelling ideas for boolean classification in genetic programming. *Genetic Programming: Proceedings of the 3rd European Conference (EuroGP 2000). Lecture Notes in Computer Science 1802*, 133–148. Springer, 2000.

[Pinker 1997] S. Pinker. *How the Mind Works*. W.W. Norton & Company. 1997.

[Poli and Langdon 1998] R. Poli and W.B. Langdon. On the search properties of different crossover operators in genetic programming. *Genetic Programming 1998: Proceedings of the 3rd Annual Conference (GP '98)*, 293–301. Morgan Kaufmann, 1998.

[Poli et al. 1999] R. Poli, J. Page and W.B. Langdon. Smooth uniform crossover, sub-machine code GP and demes: a recipe for solving high-order boolean parity problems. *Proceedings of the Genetic and Evolutionary Computation Conference (GECCO '99)*, 1162–1169. Morgan Kaufmann, 1999.

[Rana 1999] S. Rana. The distributional biases of crossover operators. *Proceedings of the Genetic and Evolutionary Computation Conference (GECCO '99)*, 549–556. Morgan Kaufmann, 1999.

[Rasheed 1999] K. Rasheed. Guided crossover: a new operator for genetic algorithm based optimization. *Proceedings of the Congress on Evolutionary Computation (CEC '99)*, 1535–1541. IEEE, 1999.

[Rosca 1996] J.P. Rosca. Generality versus size in genetic programming. *Genetic Programming 1996: Proceedings of the 1st Annual Conference*, 381–387. Morgan Kaufmann, 1996.

[Schnier and Yao 2000] T. Schnier and X. Yao. Using multiple representations in evolutionary algorithms. *Proceedings of the 2000 Congress on Evolutionary Computation (CEC '2000)*, 479–486. IEEE, 2000.

[Soule et al. 1996] T. Soule, J.A. Foster and J. Dickinson. Code growth in genetic programming. *Genetic Programming 1996: Proceedings of the 1st Annual Conference,* 215–223. Morgan Kaufmann, 1996.

[Syswerda 1989] G. Syswerda. Uniform crossover in genetic algorithms. *Proceedings of the 2nd International Conference Genetic Algorithms (ICGA '89)*, 2–9. 1989.

[Vekaria and Clack 1999] K. Vekaria and C. Clack. Biases introduced by adaptive recombination operators. *Proceedings of the Genetic and Evolutionary Computation Conference (GECCO '99)*, 670–677. Morgan Kaufmann, 1999.

[Whitley 1989] D. Whitley. The GENITOR algorithm and selective pressure: why rank-based allocation of reproductive trials is best. *Proceedings of the 2nd International Conference Genetic Algorithms (ICGA '89)*, 116–121. 1989.

[Wineberg and Oppacher 1996] M. Wineberg and F. Oppacher. The benefits of computing with introns. *Genetic Programming 1996: Proceedings of the 1st Annual Conference*, 410–415. Morgan Kaufmann, 1996.

6 Genetic Algorithms for Rule Discovery

> "We actually go to the extreme of using the problem space as the working search space."
> [Janikow 1993, p. 200]

In this chapter we discuss several issues related to developing genetic algorithms (GAs) for prediction-rule discovery. The development of a GA for rule discovery involves a number of nontrivial design decisions. In this chapter we categorize these decisions into five groups, each of them discussed in a separate section, as follows.

Section 6.1 discusses issues related to individual representation. Section 6.2 discusses several task-specific "genetic" operators, i.e., operators developed specifically for a given data mining task, or a class of related data mining tasks. Sections 6.3 and 6.4 discuss task-specific population-initialization and rule-selection methods, respectively. Finally, section 6.5 discusses fitness evaluation.

In general the operators and methods discussed in this chapter were developed for the classification task (section 2.1), but they can also be used in other data mining tasks involving the discovery of prediction rules, such as the dependence modeling task (section 2.2). These operators and methods can be regarded as a form of incorporating task-specific knowledge into the GA, or adapting the GA for the discovery of prediction rules.

6.1 Individual Representation

This section is divided into three parts. Subsection 6.1.1 discusses the pros and cons of two broad approaches (Michigan and Pittsburgh) for encoding rules into a GA individual. Subsections 6.1.2 and 6.1.3 discuss how to encode a rule antecedent and a rule consequent, respectively, into a GA individual.

6.1.1 Pittsburgh vs Michigan Approach

In conventional GAs each individual corresponds to a candidate solution to a given problem. In our case the problem is to discover prediction rules. In general, we are interested in discovering a set of rules, rather than a single rule. So, how can we encode a set of rules in a GA population? In essence there are two approaches.

First, if we follow the conventional GA approach, each individual of the GA population represents a set of prediction rules, i.e., an entire candidate solution. This is called the Pittsburgh approach.

Another approach, which departures from conventional GAs, consists of having an individual represent a single rule, i.e., a part of a candidate solution. This is called the Michigan approach. In the literature the term Michigan approach is often used to refer to classifier systems [Goldberg 1989, chap. 6], which is one specific kind of evolutionary algorithm where each individual represents a rule. In this book we use the term Michigan approach in a broader sense, to refer to any kind of evolutionary algorithm for rule discovery where each individual represents a single prediction rule.

In the Michigan approach there are at least two possibilities for discovering a set of rules. The first one is straightforward. We let each run of the GA discover a single rule (the best individual produced in all generations) and simply run the GA multiple times to discover a set of rules. An obvious disadvantage of this strategy is that it is computationally expensive, requiring many GA runs. The second possibility is to design a more elaborate GA where a set of individuals – possibly the whole population – corresponds to a set of rules. In the remainder of this subsection we assume the use of this kind of "elaborate" version of the Michigan approach.

It is important to understand the advantages and disadvantages of the Pittsburgh and Michigan approaches. The first step for this understanding is to bear in mind that these two approaches cope with the problem of *rule interaction* in different ways. The problem of rule interaction consists of evaluating the quality of a rule set as a whole, rather than just evaluating the quality of each rule in an isolated manner. In other words, the set of *best rules* is not necessarily the *best set* of rules.

Let us illustrate this point with a very simple example. Figure 6.1 shows four rules, denoted R_1, R_2, R_3 and R_4. The area inside each rule represents the data instances covered by the rule, i.e., the data instances that satisfy the rule antecedent. Assume we have a measure for the quality of each rule, denoted by $Qual(R_i)$ – $i=1,...,4$. The definition of this quality measure is not important for our current discussion. What is important for now is the relative ordering (ranking) of the rules concerning their quality. In our example, suppose this ranking turns out to be: $Qual(R_1) > Qual(R_2) > Qual(R_3) > Qual(R_4)$. Now suppose we want to select a set of two rules out of the four rules shown in Figure 6.1. If we simply selected the two best rules, R_1 and R_2, we would be selecting redundant rules, since there is a large overlap between the data space covered by the two rules. In addition, there would be no selected rule covering the data space in the right-half of the graph. This lack of rule coverage could be a serious disadvantage in a task such as classification in its standard form, where we have to discover rules covering the entire data space, in order to be able to classify every possible data instance that might show up in the test set.

In practice we can relax the restriction of finding rules with 100% of coverage by adding to the discovered rule set a default rule, which will be used to classify any data instance that does not satisfy the antecedent of any of the discovered rules. This default rule simply predicts the majority class, i.e., the class most frequent in the part of the training set not covered by any discovered rule. Ideally, however, the default rule should be used to classify few data instances, since it is a very general rule.

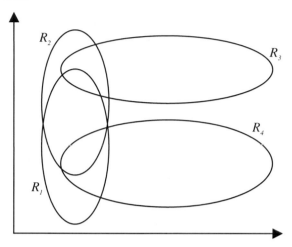

Figure 6.1: Example of rule overlapping

Now let us take a look at how the Pittsburgh and the Michigan approaches cope with the rule interaction problem. In the Pittsburgh approach, since an individual represents a rule set, we can naturally take rule interaction into account by using a fitness function that directly measures the performance of the rule set as a whole. On the other hand, in this approach an individual tends to be significantly more complex – at least, syntactically longer – than in the Michigan approach. This often leads to more complex genetic operators. In addition, the fact that at any given generation there are many more rules to be matched against the data being mined, in comparison with the Michigan approach, usually leads to a longer GA run time.

In contrast, in the Michigan approach, since an individual represents a single rule, it is syntactically shorter and the genetic operators can be simpler – by comparison with the Pittsburgh approach. The main problem with the Michigan approach is that in general the fitness function will measure the performance of a rule (individual) out of the context of the other rules, ignoring the important issue of rule interaction. Hence, assuming we want to discover a set of rules, in the Michigan approach we often need to add to the GA some method(s) that foster the discovery of a good set of rules, rather than a single good rule. In particular, we should avoid that the whole population converges to the same individual, which usually occurs in GAs. One form of preventing this undesirable convergence is to use some niching method. For instance, fitness sharing (section 5.4) was successfully used in a GA for predicting rare events [Weiss 1999]. Other ways of fostering population diversity in rule discovery will be discussed in section 6.4.

A further discussion on the pros and cons of the Michigan and Pittsburgh approaches can be found in [Freitas 2002; Greene and Smith 1993; Janikow 1993]. The latter two references also present a historical overview of some early GAs for rule discovery. In any case, hybrid Michigan/Pittsburgh systems are certainly possible. In particular, [Hekanaho 1995] proposes a GA where the setting of some parameters can be varied to explore different combinations of the Michigan and Pittsburgh approaches.

6.1.2   Encoding a Rule Antecedent

Throughout this subsection we will refer to the encoding of a single rule into an individual, so that we are implicitly assuming the use of a Michigan approach. The discussion of this section can be generalized for the Pittsburgh approach, where an individual encodes multiple rules, assuming that all rules of the individual will use the same encoding mechanism.

We start discussing how to encode rule conditions by themselves and then we address the more complex problem of encoding an entire rule antecedent – a variable-length conjunction of rule conditions.

6.1.2.1   Encoding Rule Conditions: Binary vs High-Level Encoding

When using a binary (low-level) encoding, in general each predictor attribute $Attr_i$ will be assigned a certain number of bits, which depends on the data type of the attribute. Let us consider two broad data types, categorical (nominal) and continuous (real-valued) attributes.

In the case of categorical attributes, one possibility is to encode each value of each attribute as a single bit. Using this scheme, a categorical attribute is assigned $|Dom(Attr_i)|$ bits, where $Dom(Attr_i)$ denotes the domain of $Attr_i$ – specifying the set of values that can be taken on by $Attr_i$ – and $|x|$ denotes the number of elements of set x. For each bit of $Attr_i$, if the bit is set ("1") the corresponding attribute value is included in the rule condition; otherwise that value is not included in the rule condition. Some GAs for rule discovery that use this kind of binary encoding can be found in [DeJong et al. 1993; Mansilla et al. 1999; Neri and Giordana 1995; Hekanaho 1995]. The latter two references describe GAs that discover first-order-logic rules, whereas the former two references describe GAs that discover propositional-logic rules (subsection 1.2.1).

As an example of this kind of binary encoding for categorical attributes, assume the attribute *Marital_Status (MS)* can take on four values: *single, married, divorced,* or *widow*. We can represent these values with four bits, so that the bit pattern 1000 corresponds to the rule condition (*MS = single*), whereas the bit pattern 0110 corresponds to the rule condition (*MS = married* OR *divorced*). Rule conditions containing two or more categorical values linked by a logical OR operator, such as (*MS = married* OR *divorced*), are said to involve an internal disjunction. It is interesting to note that this kind of internal disjunction is conceptually similar to the use of the subsetting approach for grouping attribute values in decision-tree building algorithms – see subsection 3.1.1.

If all bits of an attribute are set to 1 we usually assume that the corresponding rule condition will not be included in the rule antecedent, since the value of the corresponding attribute is irrelevant to determine whether or not a data instance satisfies the rule antecedent. This point will be discussed in more detail below, when we talk about the encoding of an entire rule antecedent.

In the case of continuous attributes the situation is more complex and many alternative encoding schemes are possible. Probably the simplest solution consists of discretizing the values of the continuous attribute as a preprocessing step (as discussed in section 4.2), and then encoding the resulting intervals using the same scheme used for categorical attributes. For instance, suppose that the attribute Age is discretized into five intervals: 0-18, 19-25, 26-40, 41-70, 70-∞. Now we can encode this attribute by using five bits, assigning one bit to each interval, as explained above.

This discretization makes the encoding of a GA individual simple and uniform, since all predictor attributes would be encoded by the same scheme regardless of their original data type. On the other hand, it relies on the ability of a discretization algorithm to find good intervals. As we mentioned in subsection 4.2.1, most discretization algorithms consider only one-continuous-attribute-at-a-time, so that they do not cope well with attribute interaction. The ability to cope better with attribute interaction is precisely one of the advantages of GAs in data mining, so it is important to consider the possibility of using some alternative ways of encoding continuous values directly into the genotype of an individual.

One alternative is to use a binary encoding of a floating-point number [Back 2000]. One problem with this alternative is that the length of the genome would increase fast with the number of continuous attributes, particularly if we need to use a relatively large number of bits to represent each continuous attribute, in order to get higher precision (subsection 5.3.1). A very long individual length slows down the GA.

Instead of using a binary, low-level encoding, we can encode attribute values directly into the genome. In the case of categorical attributes, the corresponding part of the genome would directly contain one of the values of the attribute. Let us consider as an example the above-mentioned $Marital\_Status$ (MS) attribute. An individual could contain the value $divorced$ for this attribute, which would be trivially "decoded" into the rule condition ($MS = divorced$). In principle one can allow the use of the above-mentioned internal-disjunction representation as well [Janikow 1993]. For instance, one can directly encode an internal disjunction such as ($single$ OR $divorced$) into the genotype.

In practice, it seems that internal disjunctions can be more easily implemented by the above-described binary encoding, which allows the use of conventional, very simple genetic operators. An important exception is the case where the categorical attributes have a large number of values. In this case a binary encoding would have the disadvantage of considerably increasing the size of an individual.

High-level encoding seems particularly advantageous in the case of continuous (real-valued) attributes, where a binary encoding tends to be somewhat cumbersome and/or inefficient, particularly for a large number of continuous attributes, as discussed above. Using a high-level encoding, the two most common ways of encoding a continuous attribute value into a GA individuals are as follows.

One way is to directly encode a single value (threshold) of the attribute into the genome, together with a suitable comparison operator. Comparison operators typically used with continuous attributes are "≤" and ">", or their dual counterpart "≥" and "<". For instance, assuming that *Age* is a continuous attribute, one can encode the operator-threshold pair "> 25" into an individual. This operator-threshold pair would be trivially decoded into the rule condition (*Age* > 25).

Another way is to directly encode both a lower threshold and an upper threshold into an individual. This requires that two comparison operators be specified. For instance, assuming that both comparison operators are specified as "≤" for all individuals, one can encode the threshold-threshold pair "25 34" into an individual. This threshold-threshold pair would be trivially decoded into the rule condition (25 ≤ *Age* ≤ 34).

Finally, note that a hybrid low-level (binary) and high-level encoding can also be used. Indeed, such a hybrid encoding seems suitable for representing prediction rules involving both categorical and continuous attributes. More precisely, one can use a low-level, binary-coded representation for categorical attributes and a high-level, real valued-coded representation for continuous attributes. This kind of hybrid representation was used, e.g., in the GA proposed by [Kwedlo and Kretowski 1999].

6.1.2.2 *Encoding a Variable-Length Rule Antecedent*

Now that we have discussed how to encode rule conditions using both low-level and high-level representations, it is time to discuss how to encode an entire rule antecedent, consisting of a variable-length conjunction of rule conditions. This is important because in principle we do not know a priori how many conditions a rule antecedent should have. A rule discovery algorithm is supposed to be autonomous enough to decide which number of conditions is most suitable for each rule, according to the data being mined. There are at least two basic approaches to cope with this problem.

First, we can use a fixed-length genotype which is suitably mapped into a variable-length rule antecedent. Second, we can use a variable-length genotype that is directly equivalent to a variable-length rule antecedent.

Figure 6.2 illustrates the use of the first approach. In the figure an individual is encoded as a set of m conditions, where m is the number of predictor attributes. Each condition $cond_i - i = 1,...,m -$ is in turn encoded as quadruple: $<Attr_i\ Op_i\ Val_{ij}\ Active_i>$, where:

- $Attr_i$ denotes the i-th predictor attribute;
- Op_i denotes the comparison operator – e.g., "=" for categorical attributes; "≤" or ">" for continuous attributes – used in the i-th condition;

- *Val$_{ij}$* denotes the *j*-th value of the domain of *Attr$_i$*;
- *Active$_i$* is a bit used as a flag to indicate whether the *i*-th condition is active ("1") or inactive ("0"). By active we mean the condition is included in the rule antecedent represented by the individual, so that it is part of the conjunction of conditions composing the rule antecedent.

condition 1		condition *m*
Attr$_1$ *Op$_1$* *Val$_{1j}$* *Active$_1$*	· · · · ·	*Attr$_m$* *Op$_m$* *Val$_{mj}$* *Active$_m$*

Figure 6.2: Fixed-length genotype for a variable-length rule antecedent

Figure 6.3 shows an example of an individual whose genotype has the general structure of Figure 6.2. Figure 6.3(a) shows the genotype of the individual. In this very simple example the data being mined has only three attributes, so three conditions are encoded in the genotype. Figure 6.3(b) shows how that genotype is decoded into a rule antecedent. Since the second condition of the genotype has the flag *Active$_2$* set to 0, that condition is removed from the decoded rule antecedent. There is an implicit logical AND operator connecting the conditions encoded into the genotype, which is explicitly shown in the decoded rule antecedent.

(*Gender = male*) 1	(*Salary = high*) 0	(*Age > 55*) 1

(a) Genotype

(*Gender = male*) AND (*Age > 55*)
(b) Decoded rule antecedent

Figure 6.3: A genotype and its corresponding decoded rule antecedent

Note that this approach implicitly assumes a positional encoding of attributes in the genotype. In other words, in each individual the first condition refers to the first predictor attribute, the second condition refers to the second attribute, and so on. Actually, given this positional convention, in practice the name of the attribute (the element *Attr$_i$* in the above discussion) does not even need to be included in the genome itself, though we will keep including it in our figures for the sake of clarity.

In any case, it is important to note that this positional convention simplifies the action of genetic operators such as crossover. By choosing the same crossover points in both parents we can directly swap the corresponding genetic material between them without risking producing some kinds of invalid offspring, such as an individual (rule antecedent) with *Attr$_i$* duplicated.

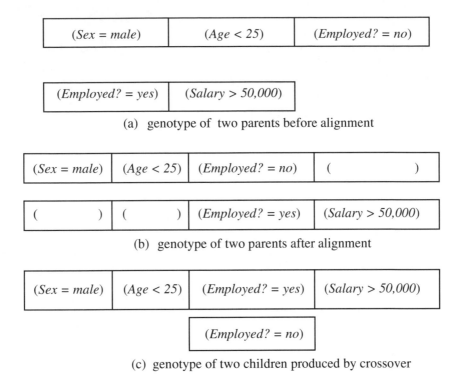

(a) genotype of two parents before alignment

(b) genotype of two parents after alignment

(c) genotype of two children produced by crossover

Figure 6.4: Crossover on variable-length individuals (rule antecedents)

Let us now turn to the second approach, which uses a variable-length genotype. Let us illustrate the use of a variable-length genotype using a high-level
encoding. Suppose that the set of predictor attributes contains 4 attributes about a
bank customer, 2 of them categorical (say *Sex, Employed?*) and 2 of them continuous (say *Age, Salary*). Suppose also that we are trying to predict whether or
not the customer has a good credit and so should or not be granted, say, a credit
card. Consider the two individuals shown in Figure 6.4(a) – each of them representing a rule antecedent. For simplicity the rule consequent is not shown. (We
will discuss some alternatives about how to encode the rule consequent in the
next subsection.)

The example shown in Figure 6.4 also assumes that the genotype implicitly
contains a logical conjunction (AND) operator between each pair of adjacent
conditions. Note that one might have a problem when performing a crossover
between the two individuals of Figure 6.4(a). For instance, if we apply a conventional one-point crossover by exchanging the rule conditions at the right of the
first condition, one of the offspring would be:

$(Employed? = yes)$ $(Age < 25)$ $(Employed? = no)$.

This is an invalid individual, since it represents a conjunctive rule antecedent
which includes two contradictory conditions: *(Employed? = yes)* and *(Employed?
= no)*.

A solution for this problem consists of first "aligning" the two individuals to undergo crossover, so that rule conditions referring to the same attribute will be located in the same position in both individuals. For instance, to perform a cross-over between the two individuals of Figure 6.4(a) one first aligns them, which could result, for instance, in the two individuals shown in Figure 6.4(b).

Now one can perform different kinds of crossover between these two individuals, simply assuming that "empty" positions produced after the alignment are part of the "genetic material" to be swapped by crossover. One possibility is to perform a conventional one-point crossover, where we first randomly choose a crossover point and then swap the genetic material to the right of that point between the two parents. For instance, in the above example choosing the crossover point after the second condition would produce the two new individuals shown in Figure 6.4(c).

Note that, in this variable-length-genotype approach, the "empty" positions of the genotype produced by alignment correspond to the "inactive" condition of the above-mentioned fixed-length-genotype approach.

In any case, one must be careful to handle some special cases. In particular, consider the case where the alignment between two parent individuals detects that they have no attribute in common. In this case crossover might produce some strange results, which may be considered invalid. For instance, consider a one-point crossover between the two aligned individuals below:

| (Sex = male) | (Age < 25) | ( ) | ( ) |

| ( ) | ( ) | (Employed? = yes) | (Salary > 50,000) |

If the crossover point happens to be between the second and third conditions, then the crossover would produce an individual representing a rule antecedent with four conditions and another individual representing an empty rule antecedent, with no condition. Such an empty rule antecedent does not make much sense. One can avoid the generation of this kind of empty rule antecedent altogether by not performing crossover when the two parent individuals have no attribute in common, or at least by forcing the crossover point to fall on a position which leads to the production of two valid offspring. Note that this special case is not restricted to the use of variable-length genotypes with alignment. It can also occur when using a fixed-length genotype (Figure 6.2), and the solutions are essentially the same: not performing crossover or choosing a crossover point that produces valid offspring.

6.1.3 Encoding a Rule Consequent

So far we have focused on the encoding of a rule antecedent, i.e., the IF part of a prediction rule, but we have said nothing about how to encode the consequent (the THEN part) of the rule. It turns out that there are several different approaches to associate a rule consequent with a given rule antecedent, as discussed in the next subsections.

6.1.3.1 Evolving the Rule Consequent

The first approach is to simply encode the rule consequent inside the genotype of an individual. Hence, in the Michigan approach one would add one more gene to that genotype. In the classification task it would be enough to associate that gene with one of the values of the goal attribute, but the name of the goal attribute itself would not need to be represented in the genome, since in this task all rules predict the same goal attribute. In another task such as dependence modeling the extra gene would have to represent both the name of the goal attribute and its predicted value. In the Pittsburgh approach one would have to add r genes to the genotype of an individual, where r is the number of rules represented in that individual. In any case, presumably the motivation for encoding a rule consequent in the genotype is that this value would be subject to the action of genetic operators and would evolve along with its corresponding rule antecedent.

We do not recommend this approach, based on the following rationale. The performance of a rule as a whole is strongly dependent on a synergistic relationship between the rule antecedent and the rule consequent. Suppose a rule antecedent has evolved to capture an important relationship in the data. It would be a shame if the rule was assigned a low fitness just because the rule consequent has not evolved yet to the "best possible" consequent for that rule. Even a small difference between the current rule consequent and the "best possible" rule consequent could greatly reduce the fitness of the rule. For instance, if a classification rule covers 20 data instances, out of which 19 belong to class c_1 and 1 belongs to class c_2, it would be terrible if the rule predicted class c_2.

6.1.3.2 Choosing the Best Rule Consequent on-the-Fly

In this approach one has a rule consequent encoded into the genotype of individuals as discussed in the previous subsection; but, instead of allowing that rule consequent to evolve along with the remainder of the genome, one has that rule consequent set by a special-purpose procedure that chooses the "best possible" consequent for the corresponding rule antecedent [Greene and Smith 1993; Noda et al. 1999]. A natural specification for this procedure would be to choose, among all possible rule consequents, the one maximizing the fitness of the rule. Presumably, in most cases there will be a relatively small number of rule consequents to choose from, since the number of goal attributes and goal attribute values tend to be small. In the classification task, for instance, there would be only C possible rule consequents, where C is the number of classes (goal-attribute values).

In this approach a new rule consequent is chosen for a rule whenever a new rule is created or modified (by some genetic operator) and its fitness is computed. To understand the basic idea of this approach, let us take a look at the example shown in Figure 6.5. This figure represents a data set for a classification task where each data instance belongs to one out of two classes, denoted "+" and "-" in the figure. The rectangle denotes the coverage of a given rule antecedent. More precisely, this rule antecedent covers 10 data instances – out of which 8 belong to the "+" class and 2 belong to the "-" class. Although we have not discussed the topic of fitness function yet, intuitively it is clear that, in order to maximize predictive accuracy, the algorithm should choose class "+" as the class predicted by the rule consequent. This approach can be regarded as a simple form of incorporating task-specific knowledge into an individual representation.

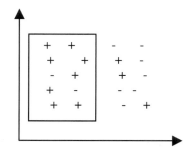

Figure 6.5: A rule antecedent covering data instances of different classes

Note that both this approach and the approach of the previous subsection have one potential problem. Since different individuals of the population can be associated with different rule consequents, some crossovers will be performed between two rule antecedents that are associated with different rule consequents. Intuitively, assuming that each of the two rule antecedents being crossed over is "optimized" for its corresponding rule consequent (which is more likely in the last stages of evolution), crossing them over seems more likely to produce offspring with lower fitness than the parents. In practice, it is not very clear how harmful this is. This is a point that deserves more research. In any case, one can avoid this potential problem either by using some kind of speciation mechanism [Deb and Spears 2000; Hekanaho 1996a] where only individuals of the same specie (in our case, having the same rule consequent) can mate with each other; or by using another way of representing rule consequents into the GA, as discussed in the next two subsections.

6.1.3.3 Using the Same Rule Consequent for the Entire Population

Another approach to represent a rule consequent in a GA is simply to associate all individuals of the population with the same rule consequent. In this case there is no need to encode the rule consequent into the genome of individuals. The same rule consequent is simply assigned to each rule antecedent when the fitness of the corresponding individual is computed.

This approach simplifies the design of the GA. In addition, since all individuals represent rules with the same rule consequent, the above-mentioned potential problem of crossover between two rule antecedents "optimized" for two different rule consequents does not occur.

This approach is particularly natural when the user is not interested in a complete classification rule set (where different rules predict different classes), but is rather only interested on rules predicting a predetermined class [Iglesia et al. 1996].

On the other hand, if the user wants to discover a set of rules with different predictions we usually have to run the GA multiple times, at least once for each rule consequent to be predicted, which can be computationally expensive. We discuss in the following two ways of implementing this approach.

First, each run of the GA can be independent of the others. This approach is very simple to implement but it has the disadvantage that it does not help to control redundancy and overlapping in the discovered rules.

Second, each run of the GA can discover rules subject to some constraint imposed by the rules discovered by previous runs of the GA. For instance, suppose we have a classification problem with two classes, c_1 and c_2. We can first run a GA to discover a rule predicting class c_1 using the entire training set. Then we remove the data instances belonging to class c_1 covered by the discovered rule from the training set, and run the GA again to discover another rule predicting class c_1. We keep doing this until all instances of class c_1 are covered by some discovered rule. We follow the same procedure for discovering rules predicting class c_2, starting again with the full training set and running the GA once for each rule predicting class c_2, always removing the instances of class c_2 covered by the rule discovered in the previous run of the GA. This kind of approach, which is sometimes called sequential covering, is used, e.g., in [Kwedlo and Kretowski 1998].

Finally, there is a special case in the general approach of associating all individuals of the population with the same rule consequent. This special case occurs in problems where the goal attribute to be predicted can take on only two values. (This case is not uncommon in practice. In many classification problems the goal attribute can take on only two classes.) In this case we have two options.

The first option consists of running the GA at least once for each rule consequent to be predicted, as discussed above. The second option consists of running the GA only for discovering rules predicting a given rule consequent, say predicting class c_1, in a given classification problem [DeJong et al. 1993; Janikow 1993]. When an unknown-class (test) data instance is to be classified by using the discovered rules we simply check whether the instance satisfies the antecedent

(IF part) of any of the discovered rules. If so it is assigned class c_1, otherwise it is assigned the other class, c_2, by default. In this approach we should use the GA to discover rules predicting the minority class (which is more difficult to be predicted) and let the majority class be the default class.

6.1.3.4 Using the Same Rule Consequent for Each Subpopulation

This approach is a relatively minor variant of the approach discussed in the previous subsection. It consists of dividing the population into subpopulations and associate with each subpopulation a fixed rule consequent. Now we can discover rules predicting different classes with a single run of the GA. In addition, once crossover can occur only between individuals of the same subpopulation, crossover is always done between individuals with the same rule consequent. It might be argued that this is one advantage. As discussed above, if two individuals are highly fit, but they have different rule consequents, a crossover between their rule antecedent could produce low-fitness individuals, which is avoided in this approach.

Two examples of the use of this approach can be found in [Araujo et al. 1999] and [Walter and Mohan 2000]. In the latter, although each subpopulation evolves separately from the others, rules from all subpopulations are considered in the process of matching the data. In contrast, in the former each rule is evaluated independently of the others.

6.2 Task-Specific Generalizing/Specializing Operators

6.2.1 Generalizing/Specializing Crossover

The basic idea of this special kind of crossover is to generalize or specialize a given rule, depending on whether the rule is currently too specific or too general, respectively [Giordana et al. 1994; Anglano et al. 1997].

The application of these operators assume that we have a method to detect when a candidate rule is too specific or too generic. In essence, a rule can be considered too specific when it is covering too few data instances, i.e., when too few instances satisfy the rule antecedent. Conversely, a rule can be considered too general when it is covering too many instances and a significant portion of the covered instances belongs to a class different from the class predicted by the rule (which is usually the most frequent class among the instances covered by the rule).

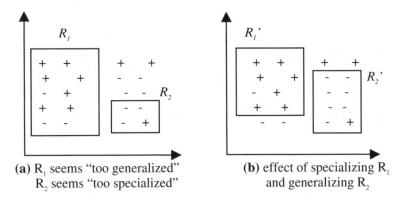

(a) R_1 seems "too generalized"   **(b)** effect of specializing R_1
R_2 seems "too specialized"   and generalizing R_2

Figure 6.6: Motivation for generalizing/specializing rules

A very simple example is shown in Figure 6.6. This figure represents a data set for a classification task where each data instance belongs to one out of two classes, denoted "+" (positive class) and "-" (negative class) in the figure. The area inside each rectangle denotes the coverage of a given rule, i.e., the data instances satisfying the corresponding rule antecedent. Figure 6.6(a) shows a situation where rule R_1 is too generalized, and so it should be specialized, and rule R_2 is too specialized, and so it should be generalized. Figure 6.6(b) shows the effect of specializing and generalizing rules R_1 and R_2, respectively, producing rules R_1' and R_2'. Of course, a rule can be generalized or specialized in several different ways. The figure shows "good" specialization and generalization operations, which intuitively seem to increase the accuracy of the rules (in the training set). In particular, in this example both rules R_1' and R_2' have a higher proportion of data instances belonging to their predicted (majority) class than their counterpart rules R_1 and R_2. This will not always be the case when carrying out a generalizing/specializing operation.

We are now ready to discuss how we can specialize or generalize a given rule by modifying the conditions in its antecedent (IF part). To simplify our discussion, assume that the evolutionary algorithm follows the Michigan approach – where each individual represents a single rule – using a binary encoding (subsection 6.1.2.1). Then the generalizing / specializing crossover operators can be implemented as the logical OR and the logical AND, respectively [Giordana et al. 1994; Anglano et al. 1997; Hekanaho 1995]. This is illustrated in Figure 6.7, where the bitwise-OR and the bitwise-AND functions are used to compute the values of the bits between the two crossover points – denoted by the "I" symbol in the Figure.

	parents		children produced by generalizing crossover			children produced by specializing crossover		
0 0	1 1 0 0	1	0 0	1 1 1 0	1	0 0	1 0 0 0	1
1 0	1 0 1 0	0	1 0	1 1 1 0	0	1 0	1 0 0 0	0

Figure 6.7: Example of generalizing/specializing two-point crossover

Although in the example of Figure 6.7 we have used two-point crossover and bitwise-OR and bitwise-AND functions to implement generalization and specialization operations, of course these operations can also be applied to other kinds of crossover and individual representation.

Consider for instance a high-level individual encoding and a conventional one-point crossover, where two children are produced by swapping the rule conditions to the right of a crossover point between two parents. Figure 6.8 illustrates the use of generalization/specialization crossover in this case. In the figure the crossover point – denoted by the symbol "I" – is located between the first and second rule conditions.

Figure 6.8(b) shows the result of applying generalizing crossover to the parents in Figure 6.8(a). For each pair of rule conditions (one from each parent) being swapped, generalization crossover implements a logical OR function between the attribute values belonging to those two conditions. We can also think of generalizing crossover as implementing, for each pair of conditions being swapped, a union of two sets of values, namely the attribute values in the corresponding rule condition of the first and second parents. The value *any* is used in the children to denote a "wild card" (or "don't care") attribute value, so that the corresponding condition is a logical tautology – it is always satisfied. In the example of Figure 6.8(b) the condition (*Employed?* = *any*) is equivalent to (*Employed?* = *no OR yes*). Similarly, the condition (*Salary* = *any*) is equivalent to (*Salary* < *40,000 OR Salary* > *30,000*). As a result, the second and third conditions of the child individuals shown in Figures 6.8(b) are effectively removed from the corresponding rules.

(Sex = male) | (Employed? = no)   (Salary < 40,000)
(Sex = female) | (Employed? = yes)   (Salary > 30,000)

(a) parents before crossover

(Sex = male) | (Employed? = any)   (Salary = any)
(Sex = female) | (Employed? = any)   (Salary = any)

(b) children produced by generalizing crossover

(Sex = male) | (Employed? = yes)   (30,000 < Salary < 40,000)
(Sex = female) | (Employed? = no)   (30,000 < Salary < 40,000)

(c) children produced by specializing crossover

Figure 6.8: Generalizing/specializing crossover in high-level individual encoding

Figure 6.8(c) shows the result of applying specializing crossover to the parents in Figure 6.8(a). For each pair of conditions (one from each parent) being swapped, specializing crossover implements a logical AND function between the attribute values belonging to those two conditions. We can also think of generalizing crossover as implementing, for each pair of conditions being swapped, an intersection of two sets of values, namely the attribute values in the corresponding rule conditions of the first and second parents.

It is interesting to note what happened with the second rule condition in the two children shown in Figure 6.8(c). If we had applied the "standard" definition of specializing crossover, both children would get the rule condition (*Employed?* = *no AND yes*). The problem is that this condition is a logical contradiction, i.e., it is never satisfied by any data instance, so that it does not make sense to have it in a rule. Since the rule antecedent is a conjunction of rule conditions, if a condition is a contradiction then its rule antecedent is never satisfied. This would certainly be a way to specialize a rule, but of course this would go too far. The rule would be so specific that it would not cover any instance, regardless of the contents of the data being mined!

To avoid this, at first glance we could simply remove the condition from the rule, producing in the genotype the condition (*Employed?* = *any*). The problem is that this operation would generalize the rule, which is exactly the opposite of what we wanted in the first place. Hence, in the example of Figure 6.8(c) we have chosen to modify the standard definition of specializing crossover, so that for each pair of conditions the specialization operation (the AND function) is applied to the corresponding attribute values only if it produces a satisfiable – not contradictory – condition. Otherwise we perform a conventional crossover for the pair of rule conditions in question. As a result, in the example of Figure 6.8(c) we ended up applying a conventional crossover on the second condition and an AND function on the third condition.

6.2.2 Generalizing/Specializing Mutation

The basic idea of generalizing/specializing a rule (see the previous subsection) is by no means restricted to the crossover operator. One can also use a mutation operator that generalizes or specializes a rule condition. (To recall the basic idea of generalizing/specializing rule conditions, see subsection 1.3.1.)

Suppose, for instance, that a given individual represents a rule antecedent with two attribute-value conditions, as follows:

(*Age* > *30*) (*Marital_Status* = *married*).

As usual, we assume that there is a logical AND operator implicit between each pair of adjacent conditions, so that the above individual actually represents a conjunction of two rule conditions. One can generalize the first condition of the above individual by using a mutation operator that subtracts a small, randomly-generated value from *30*. This might transform the above individual into, say, the following one:

(*Age* > *25*) (*Marital_Status* = *married*).

This mutated individual represents a rule antecedent that tends to cover more data instances than the original one, which is precisely the desired effect in the case of a generalization operator. The second condition of the above individual can be generalized by adding another attribute value to the condition. This might transform the above individual into, say, the following one:

$(Age > 30)$ $(Marital\_Status = married$ OR $divorced)$.

Conversely, this individual can be specialized by applying operations that are the dual of their generalization counterpart. For example, a specialization operation could transform this individual's second condition into $(Marital\_Status = divorced)$, removing the internal disjunction of that condition. As another example, if the condition with attribute Age involved a pair of threshold values, such as $(25 \leq Age \leq 35)$, that condition could be specialized by shrinking the corresponding numeric interval, producing for instance the condition $(27 \leq Age \leq 33)$.

The generalizing/specializing mutation operators described in this subsection, as well as some variants of them, can be found, e.g., in [Janikow 1993; Venturini 1993; Augier et al. 1995; Liu and Kwok 2000].

6.2.3 Condition Removal/Insertion

Another way to generalize (specialize) a rule consists of removing (inserting) a condition from (into) the rule. Condition removal can be interpreted as replacing the attribute value in the rule condition with a "don't care" ("*any*") symbol; creating, for instance, a rule antecedent of the form:

IF $(Age > 30)$ AND $(Marital\_Status = any)$, equivalent to: IF $(Age > 30)$.

This is usually called the dropping condition operator in the literature. This operator is used, e.g., in [DeJong et al. 1993; Janikow 1993; Venturini 1993]. Dropping condition generalizes a rule because the rule antecedent is a conjunction of conditions. Hence, when one removes a condition from a rule antecedent we are relaxing the criterion for a data instance to satisfy that antecedent, which tends to increase the number of data instances satisfying that antecedent.

Dropping condition is also a form of rule pruning. However, the term rule pruning usually denotes a more elaborate procedure (see subsections 3.1.2 and 3.1.3), which tries to remove irrelevant conditions from the rule, rather than just removing one condition at random.

An example of relevance-based rule pruning operator can be found in [Carvalho and Freitas 2000a]. This operator combines the probabilistic nature of genetic operators with a well-known heuristic measure of rule-condition quality called information gain [Quinlan 1993, chap. 2]. In essence this operator computes the information gain of each rule condition and removes several conditions from a rule in a probabilistic way, where the probability of a condition being removed is inversely proportional to its information gain.

Another example of relevance-based rule pruning operator can be found in [Liu and Kwok 2000]. The basic idea of this operator is to estimate the relevance of each attribute based on how well its values distinguish among data instances that are similar (nearest neighbors) to each other. An attribute is relevant if it has the same value for neighbors from the same class and different values for neighbors from different classes. This operator is also probabilistic, since the probability of a condition being removed is inversely proportional to the relevance of the attribute in that condition.

The opposite operation of condition removal is condition insertion, which consists of inserting a new rule condition into the rule. In general the new condition is randomly generated. Condition insertion specializes a rule. The reason for this is the dual of the reason why removing a condition generalizes the rule (see above).

6.2.4 Rule Insertion/Removal

If the GA follows the Pittsburgh approach, where each individual represents a rule set, there is yet another way of generalizing/specializing an individual. An individual can be generalized by adding one more rule to it. This operation generalizes a rule set because the rule set is a disjunction of rules. (We are assuming that all the rules in the set predict the same goal attribute value, i.e., they have the same consequent. Thus, a rule set is satisfied by a given data instance if any of its rule antecedents is satisfied by that instance.) Hence, when one inserts a rule into a rule set we are adding one more disjunct to that set, i.e., we are giving one more opportunity for a data instance to satisfy that rule set, which tends to increase the number of data instances satisfying that rule set.

Conversely, a rule set can be specialized by removing one rule from it. The reason why this operation specializes a rule set is the dual of the just-mentioned reason why inserting a rule generalizes a rule set.

GAs following the Pittsburgh approach that use rule insertion/removal to generalize/specialize a rule set can be found in [Janikow 1993; Kwedlo and Kretowski 1999]. In both projects rule insertion is implemented by coping a rule from one individual to another, rather than randomly generating a new rule to be inserted in the second (receiver) individual. However, the rule to be copied is randomly selected among the rules in the first (sender) individual. In the case of rule removal, the rule to be removed from an individual is also randomly selected among the rules in its rule set. Perhaps both these operators could be improved by choosing the rule to be copied or removed based on some rule-quality criterion, rather than at random. This would be conceptually similar to the idea of relevance-based rule pruning, discussed in the previous subsection.

6.2.5 Discussion

Regardless of the way in which generalization and specialization operators are implemented, it makes sense to define the probabilities of application of these two kinds of operators as proportional to the extent to which a rule is specific or generic, respectively [Giordana et al. 1994; Liu and Kwok 2000]. In other words, in this approach the probability of applying a generalization operator is increased when the individual covers few data instances, whereas the probability of applying a specialization operator is increased when the individual covers many instances having a class different from the class predicted by the rule. In this case generalizing/specializing operations can be regarded as a way of reducing the risk of overfitting and underfitting, respectively. Recall that rules that are too specific tend to overfit the training data, whereas rules that are too generic tend to underfit the training data (subsection 2.1.2).

The probability of application of generalization and specialization operators, as well as other genetic operators, can also depend on other factors related to the quality of individuals. For instance, [Liu and Kwov 2000] propose that high-fitness individuals should have a lower chance of undergoing crossover and mutation, and should instead have a higher chance to undergo generalization and specialization operations, in order to allow a further refinement of their candidate rules.

In addition, the application of generalization and specialization operators, as well as of other genetic operators, can be adaptively evolved during a GA run. For instance, [DeJong et al. 1993] extended the genotype of individuals with two operator-control bits, each of them controlling the application of a generalization operator. Each of those generalization operators can be applied only when its corresponding control bit is set ("1"). The values of the control bits are initialized at random for each individual, and they are evolved via selection, crossover and mutation along the GA run. Therefore, the system searches for good generalization operators at the same time that it searches for good rule sets.

We emphasize that the use of generalizing/specializing operators is a way to incorporate task-dependent knowledge into a GA. By task-dependent knowledge we mean knowledge about a given data mining task or a group of related data mining tasks. In this case, these operators incorporate the knowledge that in a prediction task – such as classification and dependence modeling – a rule should not be too specialized nor too generalized. Furthermore, when using these operators there can be another piece of knowledge incorporated into the GA, namely a heuristic measure of how specific or generic a candidate rule is. (Recall that this measure is necessary in order for the system to decide when to generalize or specialize a candidate rule.)

Finally, note that although knowledge about the role of generalization and specialization is dependent of the nature of the data mining task, it is independent of the application domain. Hence, generalization/specialization operators are useful for prediction-related tasks regardless of the kind of data being mined – e.g., financial data, medical data, etc.

It should be pointed out, however, that in some cases generalizing/specializing operators are simply applied in a random way, without taking into account the number of data instances covered by the individual or another heuristic information. In these cases one can say that there is less task-dependent knowledge incorporated into the GA, by comparison with the above-mentioned heuristics-based application of generalizing/ specializing operators.

6.3 Task-Specific Population Initialization and Seeding

In many applications of GAs the initial population is randomly generated. In the context of rule discovery this approach has one drawback. If we generate prediction rules at random it is possible that the generated rules will cover no training data instance, so having extremely low fitness. The problem is particularly serious if we generate long rules, i.e., rules with many conditions, since in general the larger the number of conditions in a (conjunctive) rule antecedent, the smaller the number of data instances covered by the rule.

A simple solution for this problem is to randomly initialize the population with only general rules, having a small number of conditions – say 2 or 3 conditions. A minor variant, which is somewhat more flexible, is to allow the number of conditions to vary in a wider range, but to bias the rule-generalization procedure so that short rules are more likely to be generated than long rules. In both cases, however, in general there is no guarantee that a randomly-generated rule will cover at least one data instance.

Another kind of solution is to initialize the population only with rules that are guaranteed to cover at least one instance. This can be done via seeding, i.e., one can randomly choose a training instance to act as a "seed" for rule generation [Venturini 1993; Augier et al. 1995; Hekanaho 1995; Kwedlo and Kretowski 1998]. An instance can be directly transformed into an extremely specialized rule, that covers only that instance, by using all m attribute values in the instance to instantiate the corresponding conditions of a long rule containing m conditions, where m is the number of predictor attributes in the data being mined.

In practice we do not want a rule covering a single instance, which would very likely be overfitting the data. Hence, the seeded rule can be generalized by removing several conditions from it – e.g., by applying several dropping condition operators to it – so that it covers a larger number of instances. Of course, sometimes this will not improve the quality of the seeded rule, since its generalization can cover several data instances having a goal-attribute value different from the one predicted by that rule.

An example of this generalized-seeded-rule approach can be found in [Walter and Mohan 2000], where a seed instance is generalized, via condition removal, until the corresponding rule has only two conditions.

The choice of seed instances is often done at random. This increases the degree of non-determinism of the GA, since different sets of seed instances can lead to the discovery of different rule sets. [Liu and Kwok 2000] propose a non-random form of choosing seed instances. Their seeding method is used in a GA that discovers classification rules, and is based on the idea that a seed should be a data instance lying in the middle of a cluster of instances belonging to the same class as the seed's class. As described by the authors, the method works "...*by computing, for each example [or data instance], the ratio of the numbers of same-class / opposite-class examples lying within a user-defined radius ρ. The one with the highest ratio will be selected as the initial seed.*" After a seed instance is chosen, it is used for generating the initial population. This is done by using a generalization operator.

It should be noted that the basic idea of seeding is not restricted to population initialization. A seeding operator can also be used as a "genetic" operator along a population's evolution. This approach is used, e.g., in the "universal suffrage" method for rule (individual) selection, which will be described in the next section.

6.4 Task-Specific Rule-Selection Methods

In general, rule selection is not a big problem in the Pittsburgh approach for rule discovery. Recall that in this approach an individual of the GA population corresponds to a rule set. Hence, we can use a conventional selection method for choosing the individuals (entire sets of rules) to be reproduced. After all, the fitness function can evaluate an individual's rule set as a whole, by taking into account rule interactions. This allows us to design fitness function favoring rule sets where not only each rule has a good quality but also the rules interact with each other in a synergistic way. For instance, the fitness function can (explicitly or implicitly) favor rule sets with a large degree of diversity, in the sense that different rules cover different data instances, to avoid redundant rules.

However, the situation is more complex in the case of the Michigan approach. Recall that in this approach each individual represents a single rule. Since we are usually interested in discovering a rule set, a candidate solution is usually represented by a set of individuals. This raises a question: how do we decide which rules (individuals) of the current population are selected for reproduction?

One answer for this question consists of running a procedure like the one shown in Algorithm 6.1, based on a rule-selection method proposed by [Greene and Smith 1993].

Compute Fitness of each rule (individual);
(assume that the higher the fitness, the better the rule)
Sort rules in decreasing order of Fitness;
Store the sorted rules into CandidateRuleList;
WHILE (CandidateRuleList is not empty) AND (TrainingSet is not empty)
    Remove from the TrainingSet the data instances
       correctly covered by the first rule in CandidateRuleList;
    Remove the first rule from CandidateRuleList
       and insert it into SelectedRuleList;
ENDWHILE

Algorithm 6.1: Procedure for Rule Selection Based on Data Coverage

The basic idea of this rule-selection method is to consider the data instances in the training set as "resources" of the environment being exploited by the GA individuals. Once a resource is consumed by an individual it is no longer available to other individuals. Hence, individuals compete to consume as many resources as possible, and different individuals naturally occupy different "niches" of the environment, which in our case consists of covering different subsets of data instances.

Note that the output of the above rule-selection method is a variable-length rule set. Therefore, the GA has a variable-size population. In general, the more complex a data set the higher the number of rules necessary to cover it, and so the higher the number of rules (individuals) selected to achieve this coverage. The basic idea of this method is quite simple and intuitively appealing.

Note that Algorithm 6.1 assumes that we want total coverage of the training set, in the sense that we keep selecting rules while there is at least one data instance in the training set and at least one rule in CandidateRuleList. However, when the number of data instances left in the training set is small, selecting a rule to cover the few remaining data instances can lead to overfitting. In addition, rules at the end of the CandidateRuleList tend to be low-quality rules.

A variation of this rule-selection method, which would render it somewhat more flexible, would be to relax the stopping criteria of the WHILE loop, so that in some cases one might abandon the loop earlier, without achieving total coverage of the training set. Of course, if one does abandon the loop before achieving total coverage, one will end up with some regions of the data space which are not covered by any rule. In a standard classification task this would be a problem, since we are required to classify every possible data instance to appear in the future. One solution to this problem would be simply to use a "default" rule to classify data instances not covered by any rule. This default rule would predict the majority class, i.e., the class most frequent among the training data instances that are not covered by any rule. Ideally, however, this default rule should be used to classify a small number of data instances, since it is a very general rule.

Another rule-selection method was proposed by [Giordana et al. 1994]. Like the above-described method, this one was also developed for a GA following the Michigan approach, where each individual represents a single classification rule.

This method is called "universal suffrage". The basic idea of the method, which can be naturally explained using a political metaphor, is that individuals to be mated are "elected" by training data instances. Each instance "votes" for a rule that covers it in a stochastic (probabilistic), fitness-based fashion. More precisely, let R be the set of rules (individuals) that cover a given data instance i, i.e., the set of rules whose antecedent is satisfied by instance i. Then instance i votes in one of the rules in R by using a roulette-wheel mechanism (see section 5.2). That is, each rule r in R is assigned a roulette-wheel slot whose size is proportional to the ratio of the fitness of r divided by the sum of fitness of all rules in R. Therefore, the better the fitness of r, the higher its probability of being selected over other rules covering instance i.

If there is no rule covering instance i, then the method creates a new individual (rule) covering i by using a seeding operator (see the previous section).

In the original version of the method the instances that vote are randomly-selected (at each generation), with replacement, from the entire set of training instances.

Note that, in both the above-described rule-selection methods, in general only rules covering the same instances compete with each other. Hence, these methods effectively implement a form of niching, fostering the evolution of several different rules, each of them covering a different part of the data space. In other words, they help to avoid the convergence of the population to a single individual (rule), and so can be used to make the GA discover a set of rules, rather than just one rule.

In contrast with general-purpose niching methods, the two rule-selection methods described in this section can be considered as task-specific. They were developed specifically for discovering classification rules – though they can be used in related tasks involving the discovering of prediction rules, such as dependence modeling (section 2.2).

A further discussion about niching methods in the context of rule discovery can be found in [Hekanaho 1996b] and [Dhar et al. 2000].

6.5 Fitness Evaluation

We said in subsection 1.1.1 and section 2.3 that the knowledge discovered by a data mining algorithm should satisfy three main criteria, namely: accuracy, comprehensibility and interestingness. We now discuss how these criteria can be incorporated into the fitness function of a GA.

We will assume throughout this section that the GA aims at discovering prediction rules, in particular classification rules. Fitness functions for the task of clustering will be discussed in section 8.4, and fitness functions for the data preparation task of attribute selection will be discussed in subsection 9.1.2.

The performance of a classification rule with respect to predictive accuracy can be summarized by a confusion matrix [Hand 1997; Weiss and Kulikowski 1991].

Let us start with the simplest case, namely a classification problem where there are only two classes to be predicted, referred to as the "positive" ("+") class and the "negative" ("-") class. In this case the confusion matrix will be a 2x2 matrix as illustrated in Figure 6.9. In this Figure c denotes the class predicted by a rule, which is also called the positive class. The class predicted for a given instance is c if that instance satisfies the rule antecedent, denoted by A. Otherwise the instance is assigned the negative class.

The labels in each quadrant of the matrix have the following meaning:

TP (True Positives) = Number of instances satisfying A and having class c
FP (False Positives) = Number of instances satisfying A but not having class c
FN (False Negatives) = Number of instances not satisfying A but having class c
TN (True Negatives) = Number of instances not satisfying A nor having class c

actual class

		c	not c
predicted	c	TP	FP
class	not c	FN	TN

Figure 6.9: Confusion matrix for a classification rule

Intuitively, the higher the values of TP and TN, and the lower the values of FP and FN, the better the corresponding classification rule. However, there are several different ways to measure the quality of a classification rule by using the above four variables (TP, FP, FN, TN). Before we discuss this issue in some detail, let us just say a few more words about the meaning of a confusion matrix.

So far we have assumed that this matrix contains information about the performance of a single classification rule. However, the system can also construct a confusion matrix containing information about the performance of a set of rules. That depends mainly on whether the GA follows the Michigan or the Pittsburgh approach (subsection 6.1.1), respectively. In the following we will continue to assume the use of the Michigan approach (where one individual represents a single rule), for the sake of simplicity, but the main arguments of this discussion of this section can be generalized to the Pittsburgh approach (where an individual represents a rule set) as well.

In addition, so far we have assumed that there are only two classes, producing a 2×2 confusion matrix. More generally, if there are C classes one would have a $C \times C$ confusion matrix. To simplify the evaluation of discovered rules, when there are more than two classes one can still work with 2×2 confusion matrices, as long as one evaluates one rule at a time. In this case the class c predicted by the rule is considered as the positive class, and all other classes are simply considered as negative classes, with no need to distinguish between different kinds of negative class.

Given the values of TP, FP, FN and TN, as discussed above, several rule-quality measures can be defined, as follows:

- *Precision* = TP / (TP + FP).
- *True Positive Rate* = TP / (TP + FN).
- *True Negative Rate* = TN / (TN + FP).
- *Accuracy Rate* = (TP + TN) / (TP + FP + FN + TN).

The above measures are also known by other names in the vast literature related to classification. *Precision* is sometimes called *consistency* or *confidence*. *True positive rate* is sometimes called *completeness, recall*, or *sensitivity*. *True negative rate* is sometimes called *specificity*. The terms *consistency* and *completeness* are particularly used in machine learning; the terms *precision* and *recall* are particularly used in information retrieval; and the terms *sensitivity* and *specificity* are particularly used in medical diagnosis applications.

In principle a fitness function could use any combination of the above measures to evaluate rule quality with respect to predictive accuracy. However, some
combinations seem more sensible than others. Let us start with the very simple
approach of using a fitness function based on *precision* only. This approach
seems very sensitive to overfitting problems (subsection 2.1.2). For instance,
consider a rule that covers a single data instance, and suppose that instance has
the class predicted by the rule. In this case FP = 1 and FN = 0, so that *precision* =
1 / (1 + 0) = 1 (100%), i.e., the rule has maximum *precision*. However, this rule is
probably overfitting the data, since it has no generality.

Intuitively, a fitness function that uses three or four of the variables TP, FP,
FN and TN will have the advantage of using more information.

At first glance an effective approach would be to use a fitness function based
on *accuracy rate*, which explicitly uses the four variables TP, FP, FN and TN.
However, this approach also has some problems, particularly when the class distribution is very unbalanced. For example, suppose there are 100 data instances of
the positive class and 900 instances of the negative class, so that TP + FN = 100
and FP + TN = 900. Suppose that the rule being evaluated has TP = 10, FP = 40,
FN = 90 and TN = 860. The rule's *accuracy rate* is (10 + 860) / (10 + 40 + 90 +
860) = 0.87. The rule seems to have a good predictive accuracy, but this is not the
case. The rule has a *precision* of only 10 / (10 + 40) = 0.2 and a *true positive rate*
of only 10 / (10 + 90) = 0.1. The rule seems to have a good predictive accuracy,
according to the *accuracy rate* measure, simply because the vast majority (90%)
of the data instances have a negative class, and most of the negative class instances are not covered by the rule, so that TN has a large value. Actually, this
rule has a "bad" predictive accuracy, in the sense that, instead of using the rule
for classification, one can simply use the majority rule, which predicts the majority class (in our example the negative class) for all data instances. In this example
the majority rule has an *accuracy rate* of 90%! Clearly, a better measure of predictive accuracy is necessary in problems with unbalanced classes.

There are at least two ways of evaluating a rule's predictive accuracy that
considerably mitigate some pitfalls associated with the above-mentioned problems of overfitting and unbalanced classes.

Firstly, one can use a fitness function based on the product of *precision* and
true positive rate (or *recall*), i.e.:

$$Fitness = precision \times true\_positive\_rate.$$

For instance, a fitness function based on *precision* and *true positive rate* (*recall*) was used in a GA for rare-event discovery [Weiss 1999]. Note that a rule
that is too specific (covers too few instances) will tend to have a high *precision*
but a low *true positive rate*, so that its overall fitness value will not be very high.

Secondly, one can use a fitness function based on the product of *true positive
rate* and *true negative rate* (or *sensitivity* and *specificity*), i.e.:

$$Fitness = true\_positive\_rate \times true\_negative\_rate.$$

For instance, a fitness function based on *sensitivity* and *specificity* was used in
GAs for classification-rule discovery [Carvalho and Freitas 2000a, 2000b, 2002;
Fidelis et al. 2000]. The product of *sensitivity* and *specificity* was also used as a

rule quality measure in an Ant Colony algorithm for classification-rule discovery [Parpinelli et al. 2001, 2002].

In addition, sometimes it is desirable to use application-dependent fitness functions that are not only related to predictive accuracy but also implement a more direct measure of profitability for the user. For instance, [Thomas and Sycara 1999, 2000] used a GP and a very simple GA, respectively, in a financial application where the goal was to maximize the profitability of trading rules, rather than directly maximizing the rules' predictive accuracy.

In any case, in data mining predictive accuracy is just one aspect of rule quality, as discussed in section 2.3. We now briefly discuss two other rule-quality aspects, namely rule comprehensibility and rule interestingness, which are more difficult to be effectively evaluated than predictive accuracy.

As mentioned in subsection 2.3.2, the measure of rule comprehensibility most used in the literature consists of a syntactic measure of rule or rule set length. In general, the fewer the number of rules in a rule set, the more comprehensible it is. Similarly, the fewer the number of conditions in a rule, the more comprehensible it is. When using a GA for rule discovery, this measure of rule comprehensibility can be easily incorporated into a fitness function. In general this is done by using a weighted fitness function, with a term measuring predictive accuracy and another term measuring rule comprehensibility, where each term has a user-defined weight. Variants of this basic idea are used, e.g., by [Janikow 1993; Pei et al. 1997].

This approach is very simple to implement, but it has the disadvantage of ignoring subjective aspects of rule comprehensibility. One approach that does take this subjective aspect into account consists of using a kind of interactive fitness function [Banzhaf 2000], where the user directly evaluates candidate rules represented in the individuals being evolved. One example of this approach will be discussed in subsection 9.1.2.1, in the context of attribute selection.

Finally, let us turn to the difficult issue of rule interestingness. As mentioned in subsection 2.3.3, there are two broad approaches for measuring rule interestingness, namely the objective approach and the subjective approach. One form of implementing a subjective measure of rule quality is the above-mentioned approach of interactive fitness function. Indeed, in this approach the user can evaluate a candidate rule (or rule set) with respect to both rule comprehensibility and rule interestingness at the same time.

Another kind of subjective approach consists of using an interactive GA to evolve a measure of rule interestingness. In this approach, proposed by [Williams 1999], the system works with several independent rule sets (subpopulations of rules) which are evaluated by different measures of rule interestingness. From time to time, during evolution, a small subset of rules is chosen from each population and shown to the user, for his/her subjective evaluation. The result of this evaluation seems to be used for evolving the interestingness measures – although no details about this process are given in the paper. The basic idea of this approach seems promising, but in the above reference the author did not present any computational result evaluating its performance. This kind of interactive approach has been recently used by [Poon and Prasher 2001], but this work concerns the discovery of association rules, rather than prediction rules.

Yet another kind of subjective approach consists of using a GA where the fitness function is (partially) based on the use of general impressions (see subsection 2.3.3) to measure the degree of surprisingness of a given rule. This approach was proposed by [Romao et al. 2002], where the fitness function evaluates both the predictive accuracy and the surprisingness of the rule represented by an individual. With respect to the measure of surprisingness, the basic idea is as follows. Initially, the user specifies some general impressions about the application domain. These general impressions represent the user's believes (or previous knowledge) about relationships in the data. In order to measure the degree of surprisingness associated with a given candidate rule, the GA compares the rule against the user-specified general impressions. The rule is considered surprising to the extent that it contradicts some general impression. The fitness function favors the discovery of rules that have both a high predictive accuracy and a high degree of surprisingness.

Alternatively, one can use an objective approach, where the fitness function incorporates an objective measure of rule interestingness. Again, this can be done by using a weighted fitness function, with a term measuring predictive accuracy and another term measuring rule interestingness, where each term has a user-defined weight. This approach was used by [Noda et al. 1999] in a GA specifically designed for discovering interesting prediction rules. Actually, this GA discovers rules that are interesting (according to a data-driven, objective rule interestingness measures) possibly at the expense of some small reduction in rule accuracy. (Similarly, there are several non-evolutionary rule-discovery algorithms that were explicitly designed to improve rule set simplicity possibly at the expense of some small reduction in rule set accuracy – see, e.g., [Bohanec 1994; Catlett 1991].) An objective measure of rule interestingness can also be used, among several other options of fitness function, in the GA proposed by [Flockhart and Radcliffe 1995, p. 36]. In this case it seems that the rule interestingness measure is supposed to be used alone as a fitness function, without combine it with other terms, so that there is no need for a weighted fitness function.

Although the use of weighted fitness functions is popular in the literature, it is not always a good approach. In particular, when there are several solution-quality criteria to be evaluated, it is often the case that these criteria are conflicting and/or non-commensurable, in the sense that they evaluate very different aspects of a candidate solution [Deb 2001; Fonseca and Fleming 2000]. It can be argued that this is indeed the case with some rule-quality criteria. For instance, predictive accuracy and rule comprehensibility are often conflicting criteria (improving a rule with respect to one of them can worsen the rule with respect to the other) and are in general non-commensurable criteria. Actually, the three criteria of predictive accuracy, rule comprehensibility and rule interestingness are intuitively non-commensurable, and can be conflicting in many cases.

This suggests the use of a multi-objective approach for rule discovery, where the value of the fitness function to be optimized is not a single scalar value, but rather a vector of values, where each value measures a different aspect of rule quality. In essence, a multi-objective algorithm searches for non-dominated solutions (the Paretto front). A solution S_1 is said to dominate a solution S_2 if the following two conditions hold:

(a) S_1 is better than S_2 in at least one of the objectives;
(b) S_1 is not worse than S_2 in all the objectives.

A solution S is said to be non-dominated if no solution dominates S.

A good comprehensive review of multi-objective evolutionary algorithms (MOEAs) can be found in [Deb 2001]. An overview of MOEAs can be found in [Van Veldhuizen and Lamont 2000; Fonseca and Fleming 2000]. Although there is a large body of literature about MOEAs, the use of this kind of EA for rule discovery, in the context of data mining, seems relatively underexplored. One work using a MOEA for attribute selection in the task of clustering will be discussed in subsection 9.1.5. Other projects on MOEAs for data mining and knowledge discovery can be found in [Bhattacharyya 2000a, 2000b] and [Emmanouilidis et al. 2000].

References

[Anglano et al. 1997] C. Anglano, A. Giordana, G. Lo Bello and L. Saitta. A network genetic algorithm for concept learning. *Proceedings of the 7th International Conference Genetic Algorithms (ICGA '97)*, 434–441. 1997.

[Araujo et al. 1999] D.L.A. Araujo, H.S. Lopes and A.A. Freitas. A parallel genetic algorithm for rule discovery in large databases. *Proceedings of the 1999 IEEE Systems, Man and Cybernetics Conference, v. III*, 940–945. Tokyo, 1999.

[Augier et al. 1995] S. Augier, G. Venturini and Y. Kodratoff. Learning first order logic rules with a genetic algorithm. *Proceedings of the 1st International Conference on Knowledge Discovery and Data Mining (KDD '95)*, 21–26. AAAI Press, 1995.

[Back 2000] T. Back. Binary strings. In: T. Back, D.B. Fogel and T. Michalewicz (Eds.) *Evolutionary Computation 1*, 132–135. Institute of Physics Publishing, 2000.

[Banzhaf 2000] W. Banzhaf. Interactive evolution. In: T. Back, D.B. Fogel and T. Michalewicz (Eds.) *Evolutionary Computation 1*, 228–236. Institute of Physics Publishing, 2000.

[Bhattacharyya 2000a] S. Bhattacharyya. Evolutionary algorithms in data mining: multi-objective performance modeling for direct marketing. *Proceedings of the 6th ACM SIGKDD International Conference on Knowledge Discovery and Data Mining (KDD '2000)*, 465–473. ACM, 2000.

[Bhattacharyya 2000b] S. Bhattacharyya. Multi-objective data mining using genetic algorithms. In: A.S. Wu (Ed.) *Proceedings of the 2000 Genetic and Evolutionary Computation Conference – Workshop on Data Mining with Evolutionary Algorithms*, 76–79. 2000.

[Bohanec 1994] M. Bohanec and I. Bratko. Trading accuracy for simplicity in decision trees. *Machine Learning, 15*, 223–250, 1994.

[Carvalho and Freitas 2000a] D.R. Carvalho and A.A. Freitas. A hybrid decision tree/genetic algorithm for coping with the problem of small disjuncts in data mining. *Proceedings of the Genetic and Evolutionary Computation Conference (GECCO '2000)*, 1061–1068. Morgan Kaufmann, 2000.

[Carvalho and Freitas 2000b] D.R. Carvalho and A.A. Freitas. A genetic algorithm-based solution for the problem of small disjuncts. *Principles of Data Mining and Knowledge Discovery (Proceedings of the 4th European Conference, PKDD '2000). Lecture Notes in Artificial Intelligence 1910*, 345–352. Springer, 2000.

[Carvalho and Freitas 2002] D.R. Carvalho and A.A. Freitas. A genetic algorithm with sequential niching for discovering small-disjunct rules. To appear in *Proceedings of the Genetic and Evolutionary Computation Conference (GECCO '2002)*. New York, 2002.

[Catlett 1991] J. Catlett. Overpruning large decision trees. *Proceedings of the 1991 International Joint Conference on Artificial Intelligence (IJCAI '91)*. Sidney, Australia. 1991.

[Deb 2001] K. Deb. *Multi-Objective Optimization Using Evolutionary Algorithms*. Wiley, 2001.

[Deb and Spears 2000] K. Deb and W.M. Spears. In: T. Back, D.B. Fogel and T. Michalewicz (Eds.) *Evolutionary Computation 2*, 93–100. Institute of Physics Publishing, 2000.

[De Jong et al. 1993] K. De Jong, W.M. Spears and D.F. Gordon. Using genetic algorithms for concept learning. *Machine Learning 13*, 161–188, 1993.

[Dhar et al. 2000] V. Dhar, D. Chou and F. Provost. Discovering interesting patterns for investment decision making with GLOWER – a genetic learner overlaid with entropy reduction. *Data Mining and Knowledge Discovery Journal 4(4)*, 251–280, 2000.

[Emmanouilidis et al. 2000] C. Emmanouilidis, A. Hunter and J. MacIntyre. A multiobjective evolutionary setting for feature selection and a commonality-based crossover operator. *Proceedings of the 2000 Congress on Evolutionary Computation (CEC '2000)*, 309–316. IEEE, 2000.

[Fidelis et al. 2000] M.V. Fidelis, H.S. Lopes and A.A. Freitas. Discovering comprehensible classification rules with a genetic algorithm. *Proceedings of the Congress on Evolutionary Computation – 2000 (CEC '2000)*, 805–810. La Jolla, CA, USA, IEEE, 2000.

[Flockhart and Radcliffe 1995] I.W. Flockhart and N.J. Radcliffe. GA-MINER: parallel data mining with hierarchical genetic algorithms - final report. *EPCC-AIKMS-GA-MINER-Report 1.0*. University of Edinburgh, UK, 1995.

[Fonseca and Fleming 2000] C.M. Fonseca and P.J. Fleming. Multiobjective optimization. In: T. Back, D.B. Fogel and T. Michalewicz (Eds.) *Evolutionary Computation 2*, 25–37. Institute of Physics Publishing, 2000.

[Freitas 2002] A.A. Freitas. A survey of evolutionary algorithms for data mining and knowledge discovery. To appear in: A. Ghosh and S. Tsutsui (Eds.) *Advances in Evolutionary Computation*. Springer, 2002.

[Giordana et al. 1994] A. Giordana, L. Saitta and F. Zini. Learning disjunctive concepts by means of genetic algorithms. *Proceedings of the 10th International Conference Machine Learning (ML '94)*, 96–104. Morgan Kaufmann, 1994.

[Goldberg 1989] D.E. Goldberg. *Genetic Algorithms in Search, Optimization and Machine Learning.* Addison-Wesley, 1989.

[Greene and Smith 1993] D.P. Greene and S.F. Smith. Competition-based induction of decision models from examples. *Machine Learning 13*, 229–257, 1993.

[Hand 1997] D.J. Hand. *Construction and Assessment of Classification Rules.* Wiley, 1997.

[Hekanaho 1995] J. Hekanaho. Symbiosis in multimodal concept learning. *Proceedings of the 1995 International Conference on Machine Learning (ML '95)*, 278–285.

[Hekanaho 1996a] J. Hekanaho. Background knowledge in GA-based concept learning. *Proceedings of the 13th International Conference on Machine Learning (ICML '96)*, 234–242. 1996.

[Hekanaho 1996b] J. Hekanaho. Testing different sharing methods in concept learning. *TUCS Technical Report No. 71.* Turku Centre for Computer Science, Finland. 1996.

[Iglesia et al. 1996] B. Iglesia, J.C.W. Debuse and V.J. Rayward-Smith. Discovering knowledge in commercial databases using modern heuristic techniques. *Proceedings of the 2nd International Conference on Knowledge Discovery and Data Mining (KDD '96)*, 44–49. AAAI Press, 1996.

[Janikow 1993] C.Z. Janikow. A knowledge-intensive genetic algorithm for supervised learning. *Machine Learning 13*, 189–228, 1993.

[Kwedlo and Kretowski 1998] W. Kwedlo and M. Kretowski. Discovery of decision rules from databases: an evolutionary approach. *Proceedings of the 2nd European Symposium on Principles of Data Mining and Knowledge Discovery (PKDD '98) - Lecture Notes in Computer Science 1510*, 370–378. Springer, 1998.

[Kwedlo and Kretowski 1999] W. Kwedlo and M. Kretowski. An evolutionary algorithm using multivariate discretization for decision rule induction. *Proceedings of the 3rd European Conference on Principles and Practice of Knowledge Discovery in Databases (PKDD '99) Lecture Notes in Computer Science 1704*, 392–397. Springer, 1999.

[Liu and Kwok 2000] J.J. Liu and J.T.-Y. Kwok. An extended genetic rule induction algorithm. *Proceedings of the 2000 Congress on Evolutionary Computation (CEC '2000).* IEEE, 2000.

[Mansilla et al. 1999] E.B. Mansilla, A. Mekaouche and J.M.G. Guiu. A study of a genetic classifier system based on the Pittsburgh approach on a medical domain. *Proceedings of the 12th International Conference Industrial and Engineering Applications of Artificial Intelligence and Exp. Syst. (IEA/AIE '99) - Lecture Notes in Artificial Intelligence 1611*, 175–184. Springer, 1999.

[Neri and Giordana 1995] A.Giordana and F. Neri. Search-intensive concept induction. *Evolutionary Computation 3(4)*, 375–416, 1995.

[Noda et al. 1999] E. Noda, A.A. Freitas and H.S. Lopes. Discovering interesting prediction rules with a genetic algorithm. *Proceedings of the Conference on Evolutionary Computation – 1999 (CEC '99)*, 1322–1329. Washington D.C., USA, 1999.

[Parpinelli et al. 2001] R.S. Parpinelli, H.S. Lopes and A.A. Freitas. An ant colony based system for data mining: applications to medical data. *Proceedings of the Genetic and Evolutionary Computation Conference (GECCO '2001)*, 791–798. Morgan Kaufmann, 2001.

[Parpinelli et al. 2002] R.S. Parpinelli, H.S. Lopes and A.A. Freitas. In: H.A. Abbass, R. Sarker, C. Newton (Eds.) *Data Mining: A Heuristic Approach*, 191–208. Idea Group Publishing, London, 2002.

[Pei et al. 1997] M. Pei, E.D. Goodman and W.F. Punch. Pattern discovery from data using genetic algorithms. *Proceedings of the 1st Pacific Asia Conference on Knowledge Discovery and Data Mining*. Feb. 1997.

[Poon and Prasher 2001] J. Poon and S. Prasher. A hybrid approach to support interactive data mining. *Proceedings of the Genetic and Evolutionary Computation Conference (GECCO '2001)*, 527–534. Morgan Kaufmann, 2001.

[Quinlan 1993] J.R. Quinlan. *C4.5: Programs for Machine Learning*. Morgan Kaufmann, San Mateo, 1993.

[Romao et al. 2002] W. Romao, A.A. Freitas and R.C.S. Pacheco. A genetic algorithm for discovering interesting fuzzy prediction rules: applications to science and technology data. To appear in *Proceedings of the 2002 Genetic and Evolutionary Computation Conference (GECCO '2002)*. New York, 2002.

[Thomas and Sycara 1999] J.D. Thomas and K. Sycara. In: A.A. Freitas (Ed.) *Data Mining with Evolutionary Algorithms: Research Directions – Papers from the AAAI '99/GECCO '99 Workshop. Technical Report WS-99–06*, 7–11. AAAI Press, 1999.

[Thomas and Sycara 2000] J.D. Thomas and K. Sycara. In: A.S. Wu (Ed.) *Proceedings of the 2000 Genetic and Evolutionary Computation Conference Workshop Program – Worshop on Data Mining with Evolutionary Algorithms*, 72–75. 2000.

[Van Veldhuizen and Lamont 2000] D.A. Van Veldhuizen and G.B. Lamont. Multiobjective evolutionary algorithms: analyzing the state-of-the-art. *Evolutionary Computation 8 (2)*, 125–147, 2000.

[Venturini 1993] G. Venturini. SIA: a supervised inductive algorithm with genetic search for learning attributes based concepts. *Machine Learning: Proceedings of the 1993 European Conference (ECML '93) - Lecture Notes in Artificial Intelligence 667*, 280–296. Springer, 1993.

[Walter and Mohan 2000] D. Walter and C.K. Mohan. ClaDia: a fuzzy classifier system for disease diagnosis. *Proceedings of the 2000 Congress on Evolutionary Computation (CEC '2000)*. IEEE, 2000.

[Weiss 1999] G.M. Weiss. Timeweaver: a genetic algorithm for identifying predictive patterns in sequences of events. *Proceedings of the Genetic and Evolutionary Computation Conference (GECCO '99)*, 718–725. Morgan Kaufmann, 1999.

[Weiss and Kulikowski 1991] S.M. Weiss and C.A. Kulikowski. *Computer Systems that Learn*. Morgan Kaufmann, San Mateo, 1991.

[Williams 1999] G.J. Williams. Evolutionary hot spots data mining: an architecture for exploring for interesting discoveries. *Proceedings of the 3rd Asia-Pacific Conference on Knowledge Discovery and Data Mining (PAKDD '99)*, 184–193. Springer, 1999.

7 Genetic Programming for Rule Discovery

> "...it is possible to define genetic programming
> as the direct evolution of programs or algorithms
> for the purpose of inductive learning."
> [Banzhaf et al. 1998, p. 6]

In subsection 5.4.4 we saw that standard Genetic Programming (GP) for symbolic regression – where all terminals are real-valued variables or constants and all functions have real-valued inputs and output – can be used for classification, if the numeric value output at the root of the tree is properly interpreted. However, this kind of GP does not produce high-level, comprehensible IF-THEN rules in the style of the rules discovered by rule induction and decision-tree building algorithms (chapter 3) and GAs for rule discovery (chapter 6). As discussed in subsection 1.1.1, rule comprehensibility is important whenever the discovered knowledge is used for decision making by a human user.

Therefore, in this chapter we are mainly interested in using GP for discovering high-level, comprehensible prediction (IF-THEN) rules, rather than just producing a numerical signal at the root node of a GP tree.

Section 7.1 discusses the problem of closure in GP for rule discovery. The next three sections discuss three different ways of coping with this problem, namely booleanizing all terminals (section 7.2), using a constrained-syntax or strongly-typed GP (section 7.3) and using a grammar-based GP (section 7.4). Section 7.5 discusses several different ways of using GP for building decision trees. Finally, section 7.6 discusses the quality of prediction rules discovered by GP.

7.1 The Problem of Closure in GP for Rule Discovery

One important problem to be addressed by Genetic Programming (GP) for rule discovery is the closure property of GP. As discussed in section 5.4.1, this property means that the output of a node in a GP tree can be used as the input to any parent node in the tree. Note that this property is satisfied in the above-mentioned case of standard symbolic-regression GP applied to numeric data, using mathematical functions in the function set. In this case all tree nodes output a real-valued number that can be used as the input to any other tree node.

When mining a data set containing a mixture of continuous (real-valued) and categorical (nominal) attribute values the closure property is not satisfied by standard GP. Different attributes are associated with different operators/functions. For instance, the condition (*Age* < *18*) is valid, but the condition (*Sex* < *female*) is not. As a result, a standard GP system might generate invalid trees such as the one shown in Figure 7.1.

The tree shown in this figure illustrates two kinds of violation of data type restrictions. The rule condition (*Sex* > *male*) is an invalid propositional-logic condition, since the operator ">" does not make sense for a categorical attribute such as *Sex*. The rule condition (*Age* < *Salary*) can be considered an invalid first-order-logic condition (from a semantical viewpoint), since the attributes *Age* and *Salary* have different, incompatible domains.

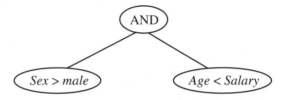

Figure 7.1: Example of invalid GP individual

Several solutions have been proposed to cope with the closure requirement of GP in the context of prediction-rule discovery. We divide the proposed solutions in three groups, namely: (a) booleanizing all terminals and using only boolean functions; (b) using a constrained-syntax or strongly-typed GP; (c) using a grammar-based GP. We discuss each of these approaches in turn in the next three sections.

7.2 Booleanizing All Terminals

Obviously, one general approach to satisfy the closure requirement is simply not to use multiple data types. That is, if the data set being mined contains attributes with different data types, we can convert all data types to a single, standard data type, and then include in the function set only functions having arguments and producing outputs of that data type.

In this section we discuss one particular approach for data type standardization that consists of booleanizing all attributes. This is probably the simplest individual-representation approach for coping with the closure property of GP in rule discovery. Prediction rules are essentially logical combinations of attribute-value conditions. Hence, the closure property is satisfied by booleanizing all terminals in the terminal set and using only boolean functions (logical AND, OR, etc.) in the function set. Now all tree nodes output a boolean value (*true* or *false*) that can be used as an input by any other tree node.

The booleanization of the terminal set is usually done as a preprocessing step for the GP. As a result of booleanization, in general each terminal in the terminal set corresponds to a boolean rule condition or to a boolean attribute. From the viewpoint of the GP, this approach consists of modifying the data being mined (assuming the original attributes are not boolean) rather than modifying a standard GP algorithm, at least from a *conceptual* point of view.

At the physical level of data storage we could keep the original data as it is, as long as we compute the proper boolean value for each terminal node of a GP tree for each data instance on-the-fly, as the predictor attributes of that data instance are input into the tree. In practice it seems more efficient to actually modify the data file, replacing the original attribute values of each data instance with precomputed boolean values for terminals in the terminal set. This tends to improve computational efficiency particularly if the terminals consist of complex conditions, since each data instance will be accessed many times – for fitness-computation purposes – during the GP evolution.

An example of this approach is illustrated in Figure 7.2. This figure shows a GP individual where all terminal nodes consist of rule conditions in the form *<Attr Op Val>*, where *Attr* is a predictor attribute, *Op* is a comparison operator in $\{=, \leq, >\}$, and *Val* is a value belonging to the domain of *Attr*. Each of these terminal nodes outputs a boolean value, *true* or *false*, depending on whether or not the corresponding condition is satisfied for a given data instance. In the figure the internal nodes are logical operators in {AND, OR}.

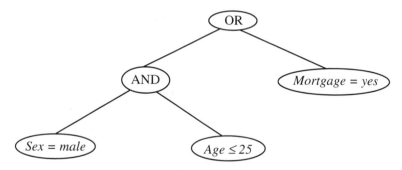

Figure 7.2: Example of GP individual having only boolean terminals

The individual in this figure represents the following antecedent (IF part) of a prediction rule:

IF ((*Sex* = *male*) AND (*Age* ≤ 25)) OR (*Mortgage* = *yes*).

The consequent (THEN part) of this rule could be determined by any of the methods discussed in subsection 6.1.3. This issue is quite orthogonal to the issues of booleanization and closure discussed here. In any case, let us assume that a single rule consequent is associated with the entire individual of Figure 7.2. In this case the individual can be regarded as a rule set in disjunctive normal form (DNF), i.e., a disjunction of rules where each rule is a conjunction of conditions. In other words, the individual effectively represents a rule set, rather than a single rule. To see this point, let c_i be the class predicted by the rule consequent. Then the above individual in DNF can be effectively decomposed into the following set of two rules:

IF (*Sex = male*) AND (*Age* ≤ 25) THEN (class = c_l)
IF (*Mortgage = yes*) THEN (class = c_l).

DNF is a simple and intuitive way of presenting prediction rules for the user. To facilitate the conversion of a GP tree into a set of rules in DNF one can use some syntactic restrictions to guarantee that, in any individual tree, an OR node is never a descendant of an AND node. For instance, for a function set consisting of the boolean operators {AND, OR}, some restrictions that facilitate the conversion of a tree into a rule set in DNF are as follows:

(a) An OR node can have as its parent only an OR node, and it can have as its children any kind of node (OR, AND, or a terminal node consisting of a rule condition);
(b) An AND node can have as its parent either an OR node or an AND node, and it can have as its children AND nodes or terminal nodes.

The use of a DNF-based individual representation also facilitates the implementation of some semantic restrictions – see subsection 7.2.3.

More generic syntactic restrictions, which can be applied to data sets with mixed categorical and continuous data types, will be discussed in section 7.3.

Note that in the example of Figure 7.2 all terminal nodes contain propositional-logic rule conditions. For the sake of simplicity, in this section we limit our discussion to the booleanization of propositional-logic rule conditions. However, the basic idea of booleanizing all terminal nodes is essentially orthogonal to the kind of rule representation used. We can also apply this idea to first-order-logic rule conditions. In this case the terminal set would also contain boolean terminals such as "$Attr_i > Attr_j$", where $Attr_i$ and $Attr_j$ are two distinct attributes having the same domain (data type) – or at least "compatible" domains.

7.2.1 Methods for Booleanizing Attributes

So far we have said nothing about how one booleanizes attributes and conditions to create the boolean terminals that will compose the terminal set. We now turn to this issue. We first discuss the case involving a categorical (nominal) attribute and then the case involving a continuous (real-valued) attribute.

Let $Attr_i$ be a categorical attribute whose domain has V_i distinct values. This kind of attribute can be booleanized in at least two different ways. The first approach consists of "partitioning" this attribute into V_i new boolean attributes, each of them defined by the condition: "$Attr_i = Val_{ij}$", $j=1,...,V_i$, where Val_{ij} is the j-th value of $Attr_i$'s domain. For instance, consider the attribute *Company_Scope* whose domain is, say, the following set of values: {*regional, national, international*}. We can transform this attribute into the following three new boolean attributes: "*Company_Scope = regional*", "*Company_Scope = national*", "*Company_Scope = international*".

Another approach to booleanize a categorical attribute consists of "partitioning" it into just two boolean attributes, regardless of the number V_i of distinct values in the domain of the attribute. In this approach the set of V_i values of the original attribute is partitioned into two mutually exclusive and exhaustive sets, one of them with U_i ($1 \leq U_i \leq V_i$ -1) values and the other one with the remaining V_i - U_i values. Each of the two just-created boolean attributes will be defined by a condition of the form "$Attr_i$ in $set\_of\_values_i$", where $set\_of\_values_i$ is a set of V_i or V_i - U_i values of $Attr_i$'s domain.

For instance, consider again the attribute *Company_Scope* whose domain is the set of values: {regional, national, international}. We can transform this attribute into one of the following three different pairs of new boolean attributes:

"*Company_Scope* in {*regional*}", "*Company_Scope* in {*national, international*}"
"*Company_Scope* in {*national*}", "*Company_Scope* in {*regional, international*}"
"*Company_Scope* in {*international*}", "*Company_Scope* in {*regional, national*}".

In this example $V_i = 3$, $U_i = 1$, and $V_i - U_i = 2$. Whenever $U_i = 1$ (or equivalently $U_i = V_i - 1$, which is the dual situation), as in this example, we can replace the "in" operator by the "=" or "≠" operator in order to get syntactically-shorter – yet semantically-equivalent – boolean-attribute definitions. For instance, in the above example the first pair of boolean-attribute definitions could be expressed as:

"*Company_Scope = regional*", "*Company_Scope ≠ regional*".

Let us now discuss the pros and cons of each of the above approaches for booleanization. The first approach, which creates V_i new boolean attributes, is straightforward to implement, but it has the disadvantage of producing several new attributes. Assuming that each new boolean attribute will be included in the terminal set of GP, this can significantly increase the search space of GP.

In contrast, the second approach produces only two new attributes, which reduces the search space for GP, but it introduces a non-trivial problem. How can we choose which of the values of the original attribute will be grouped together in each of the new boolean attributes? Clearly, this decision should not be made at random, since a bad decision could easily hinder the discovery of good prediction rules by the GP.

To see the importance of making a good decision, let us expand the above example involving the attribute *Company_Scope*, with domain {*regional, national, international*}. Suppose that we have a classification problem with two classes to be predicted: "+" and "-". To keep things very simple, suppose we have a training set with only 9 data instances, each of them with the values of *Company_Scope* and *Class* shown in Table 7.1.

In practice the data set would have many more predictor attributes, of course, but in general attributes are booleanized in an one-attribute-at-a-time basis. This approach is much faster than an approach based on many-attributes-at-a-time, since in the latter many more different combinations of attributes would have to be considered. For instance, suppose that there are 100 attributes. In the one-attribute-at-a-time approach there are 100 "combinations" of single attributes. In the two-attribute-at-a-time approach there are 100! / (98! × 2!) = 4995 combinations of two attributes. Of course, the gain in computational efficiency associated with the one-attribute-at-a-time approach is obtained at the expense of ignoring attribute interactions, which can reduce the quality of the produced boolean attributes.

Table 7.1: Example of data used as input for booleanizing a categorical attribute

Company_Scope	Class
regional	+
regional	-
regional	-
national	-
national	-
national	-
international	+
international	+
international	+

In the example of Table 7.1 a booleanization that produces the new pair of attributes:

"Company_Scope in {*regional*}", *"Company_Scope* in {*national, international*}"

clearly makes the discovery of good classification rules by the GP more difficult. The reason is that this booleanization is not very helpful for class discrimination. For instance, including the boolean attribute *"Company_Scope* in {*national, international*}"* in a rule antecedent will not be very helpful, since the distribution of classes among the examples satisfying this condition is 50%-50% (3 examples with "+" class and 3 with "-" class). Intuitively, a better booleanization would be to produce the following new pair of attributes:

"Company_Scope in {*international*}", *"Company_Scope* in {*regional, national*}".

Each of these two boolean attributes is quite useful for discriminating between the classes "+" and "-", given the data shown in Table 7.1.

Given the importance of performing a good-quality booleanization, one often uses either a heuristic or the opinion of a domain expert (if available) to choose which of the values of the original attribute will be grouped together in each of the new boolean attributes. For instance, [Hu 1998] used the well-known information gain ratio measure – see [Quinlan 1993, chap. 2] – as a general heuristic for the booleanization of attributes in several different data sets, while [Bojarczuk et al. 1999] used the help of a domain expert for booleanizing attributes of a data set concerning chest pain diagnosis.

Finally, we turn to the problem of booleanizing continuous (real-valued) attributes. Conceptually speaking the problem is similar to the booleanization of categorical attributes. One difference is that in the case of continuous attributes the first approach mentioned above, namely creating one new boolean attribute for each value of the original attribute, does not make sense, given the nature of continuous attributes.

In the case of continuous attributes the most common approach is to partition the set of original attribute values into two mutually exclusive and exhaustive subsets, each of them assigned to a new boolean attribute. This is conceptually similar to the corresponding approach for booleanizing categorical attributes.

Indeed, in order to choose which of the values of the original continuous attribute will be grouped together in each of the new boolean attributes, one can use basically the same heuristic used for booleanizing categorical attributes. (See, e.g., [Quinlan 1993, chap. 2] for an adaptation of the information gain measure for continuous attributes.) This is the approach used in [Hu 1998].

As an alternative approach, the set of original attribute values could be partitioned into a small number κ ($\kappa > 2$) of mutually exclusive and exhaustive subsets, each of them assigned to a new boolean attribute. This approach is less used, due mainly to its being more complex and more computationally expensive than the binary partition (where $\kappa = 2$).

7.2.2 Examples of GP Systems Booleanizing Terminals

Table 7.2 lists some GP systems using the approach of booleanizing all terminals to satisfy the closure property of GP. The first column of this table mentions the reference(s) where the system is described. The second column indicates the kind of data mining task being solved by the system.

Most systems mentioned in this table have been developed for discovering classification rules. Hu's system addresses the task of attribute construction as a preprocessing step for classification. This system will be discussed in subsection 9.4.1. The third column indicates the function set used by each system. All systems use the logical AND and OR operators, and some systems also use additional logical operators.

Table 7.2: Examples of GP systems that booleanize all terminals

GP system	data mining task	function set
[Bojarczuk et al. 1999, 2000]	discovery of classification rules	{AND, OR, NOT}
[Eggermont et al. 1999a, 1999b]	discovery of classification rules	{AND, OR, NAND, NOR}
[Hu 1998]	attribute construction for classification	{AND, OR}

7.2.3 A Note on Semantic Constraints

Although booleanizing all terminals meets the closure requirement and guarantees the generation of syntactically-valid trees, we still need to be careful to avoid the generation of semantically-invalid trees such as the tree shown in Figure 7.3.

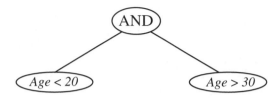

Figure 7.3: Semantically-illegal GP individual

In the figure we assume that a GP individual is represented by a tree that specifies the set of attribute-value conditions of a given rule antecedent. This tree is semantically invalid because it corresponds to a rule antecedent where both the condition (*Age < 20*) and the condition (*Age > 30*) have to be satisfied by a data instance, which is clearly impossible. In other words, this rule antecedent would not cover any data instance, and so it would be useless.

Although a suitable fitness function will naturally do the job of removing this kind of semantically-invalid individual from the population, in general it is more cost-effective to avoid the generation of this kind of individual in the first place. This can be done by incorporating appropriate restrictions into the generation of the initial population and into genetic operators that create new individuals.

For instance, crossover and mutation operations can be extended with repair procedures that remove all but one of mutually-inconsistent conditions. In the example of Figure 7.3, either the condition (*Age < 20*) or the condition (*Age > 30*) would be removed by such a repair operator. This removal could be implemented by replacing the removed condition with a new randomly-generated condition.

Alternatively, if a given crossover would produce a semantically-invalid individual the algorithm can refuse to apply that particular instance of crossover. In this case it can either try to perform the crossover again with the same parents but with different crossover points or simply have the crossover return the original parents.

In any case, the basic idea is to enforce the semantic restriction that there cannot be more than one terminal node referring to the same attribute in the same rule antecedent. (Note that if an individual represents a rule set, it is easier to enforce this restriction if the individual's genotype is in disjunctive normal form [Mendes et al. 2001], as discussed above.)

7.3 Constrained-Syntax and Strongly-Typed GP

One approach to go around the closure requirement of GP, being able to manipulate several different data types without producing invalid individuals, consists of using some form of constrained-syntax or strongly-typed GP.

The key idea of constrained-syntax GP is that, for each function *f* available in the function set, the user specifies which terminals/functions can be used as children nodes for a node containing function *f* [Koza 1992].

[Montana 1995] proposed a relatively-minor variation of this approach, which he calls "basic Strongly-Typed GP (STGP)". Basic STGP specifies which children each function node can have in an indirect way. For each function f available in the function set, the user specifies only the data type of its arguments and the data type of its result, rather than directly specifying which terminals/functions can be used as children nodes for a node containing function f. In addition, each terminal is also assigned a specific data type. Each function node in the GP tree can have only children nodes with the data types specified for the arguments of that function.

In other words, in order to generate only syntactically-legal trees, basic STGP essentially enforces the following rules: (a) the root node of the tree returns a value of the data type required by the target problem; and (b) each nonroot node returns a value of the data type required by its parent node as an argument.

This is the approach followed, e.g., by [Bhattacharyya et al. 1998; Bojarczuk et al. 2001]. An example of this approach is shown in Table 7.3, where, for each row, the second and third columns specify the data type of the input and the output of the function specified in the first column. Once this kind of specification is available to the GP, the system can generate individuals such as the one shown in Figure 7.4. This individual represents the rule antecedent:

$$IF\ (Attr_6 = yes)\ AND\ (\ (Attr_9 \times 2.7) < (Attr_1 + Attr_4)\ ).$$

The consequent (THEN part) of this rule could be determined by any of the methods discussed in subsection 6.1.3. This issue is quite orthogonal to the issue of strongly-typed GP discussed here.

Table 7.3: Example of data type definitions for input and output of functions

Functions	data type of input arguments	data type of output
+, -, *, /	(real, real)	real
≤, >	(real, real)	boolean
=	(nominal, nominal)	boolean
AND, OR	(boolean, boolean)	boolean

Syntactic constraints are normally enforced during the entire GP run, whenever a new individual is created – when the initial population is generated and when a new individual is produced by some genetic operator such as crossover or mutation. In other words, the population-initialization procedure and genetic operators such as crossover and mutation are modified to create only valid trees, by respecting the user-defined restrictions on tree syntax.

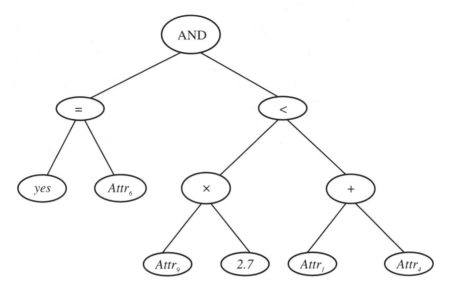

Figure 7.4: A GP individual meeting the data type restrictions of Table 7.3

In particular, syntactically-constrained crossover generally works as follows. The crossover point of the first individual is randomly chosen. Then the crossover point of the second individual is chosen among the tree nodes having the same data type as the first individual's crossover point. If no node of the second individual has the required data type, we can have the crossover operator return the parent individuals or, alternatively, try again to perform crossover with the same parents, this time choosing a different data type for the crossover point of the first individual. (As long as the number of different data types is not large, the overhead of trying more than one crossover for a given pair of individuals tends to be relatively small, especially considering that the majority of GP processing time is typically taken by fitness evaluation, rather than by genetic operators.)

The above-described basic ideas of constrained-syntax GP were also used in a GP developed for mining an object-oriented database, where there are several different data types (object classes) [Ryu and Eick 1996]. In this system the GP searches for an object-oriented database query that describes the commonalities in a predetermined set of objects.

[Montana 1995] also introduced a more elaborate form of STGP, involving generic functions. Actually, generic functions are not strongly-typed functions but rather classes of strongly-typed functions.

A generic function is a function that can take a variety of different argument data types and, in general, return values of a variety of different types. The data type of the value returned by a generic function is exactly determined by the data type of all the arguments of the function. Specifying the data type for each of the arguments of a generic function (and hence determining its return data type) is called "instantiating" the generic function.

To be included in the tree representing an individual of the GP population a generic function must be instantiated, which is done when a new tree or subtree having a generic function is generated – when creating the initial population or when performing a mutation, respectively. Once instantiated the function keeps the same argument data types whenever it is passed from a parent individual to a child one. However, note that there can be different instantiations of a generic function in a given individual.

The main motivation for using generic functions is that it allows us to define a single function that can perform the same basic operation in a number of different situations, where each situation is characterized by a given set of data types specified for the arguments of the function. For instance, a generic ADD function can be used to add any two attributes of the same data type, such as

> "*Salary* ADD *Current_Account_Balance*" or
> "*Number_of_years_in_job1* ADD *Number_of_years_in_job2*".

In the former case the attributes could be considered to belong to the data type *money*, while in the latter case the attributes could be considered to belong to the data type *number_of_years*.

To the best of our knowledge the use of generic functions in prediction-rule discovery is very little explored in the literature. This is probably mainly due to the fact that in most cases (at least in the academic literature) the attributes being mined are assigned only very coarse-grained data types, such as a *continuous* or *categorical* data type. In practice, however, finer-grained data types can be associated with the attributes being mined. For instance, in the above example *money* and *number_of_years* can be considered finer-grained data types, being "particular cases" of the coarser-grained data type *continuous*.

In general the use of finer-grained data types would create more need and more opportunities for using generic functions. It seems that this would have the benefit of incorporating more "semantic" restrictions into the data mining algorithm.

Intuitively, the use of finer-grained data types would be particularly useful for discovering rule conditions expressed in first-order logic – conditions of the form ($Attr_i$ Op $Attr_j$), where $Attr_i$ and $Attr_j$ are two distinct attributes with the same (or compatible) data type and Op is a comparison operator (section 1.2.1). In this case the use of finer-grained data types naturally avoids the generation of invalid rule conditions such as (*Salary* > *Number_of_years_in_job1*). If these two attributes were instead assigned a coarse-grained data type such as *continuous* then the attribute typing system would not prevent the GP from generating such an invalid rule condition.

Actually, in a real-world database system attributes often have (or should have) more elaborate, finer-grain domains (data types). This kind of information can be used to discover first-order-logic prediction rules whose antecedents contain only compatible attributes, as discussed, e.g., in [Freitas 1998]. This work proposes to explicitly compute an Attribute-Compatibility Table (ACT), based on information about attribute domains retrieved from the database being mined. An ACT is essentially a two-dimensional matrix with m rows and m columns, where m is the number of attributes being mined. If attributes A_i and A_j – i,j in [1..m] –

are compatible, the corresponding entry *i,j* of the ACT has the value "yes", otherwise that entry has the value "no". This kind of ACT is then used not only to generate individuals of the initial population but also to apply genetic operators during evolution, in order to guarantee that only valid individuals (with respect to data type combinations in a rule antecedent) are generated.

Another possibility to incorporate semantic restrictions into a GP system is to make it aware of the dimensional units – e.g., meter, second, etc. – of some attributes. Hence the GP would be able to take this semantic information into account when combining different attributes into the same rule antecedent. This approach, called dimensionally-aware GP, has recently been proposed by [Keijzer and Babovic 1999]. It seems particularly useful in the discovery of scientific laws, where attributes are likely to be associated with well-defined dimensional units.

7.4 Grammar-Based GP for Rule Discovery

Another approach to go around the closure requirement of GP, being able to manipulate several different data types without producing invalid individuals, consists of using a grammar to enforce syntactic and semantic constraints. This approach can be considered a variant of the constrained-syntax and strongly-typed approaches described in section 7.3. In this variant the syntactic and semantic constraints are specified by a grammar.

We emphasize that the grammar can be used not only to enforce syntactic restrictions but also to enforce elaborate semantic restrictions based on the domain knowledge provided by a domain expert.

A grammar-based GP for rule discovery is described in detail in a recent book by [Wong and Leung 2000]. One of the case studies described in this book illustrates well how one can use a grammar to encode domain-specific, user-specified background knowledge. We summarize below this case study.

The data being mined is a fracture data set containing 6,500 records and 8 attributes, namely: *Sex, Age, Admission Date (AdmDay), Length of Stay in Hospital (Stay), Diagnosis of Fracture (Diagnosis), Operation, Surgeon, Side of Fracture (Side)*. A medical expert suggested that the set of 8 attributes could be divided into groups of attributes corresponding to the following 3 temporally-ordered events: Firstly, a diagnosis is given to the patient. The attribute *Diagnosis* depends on the attributes *Sex, Age* and *AdmDay*. Secondly, an operation is performed. The attribute *Operation* depends on the previous three attributes (*Sex, Age* and *AdmDay*) and on the attribute *Diagnosis*. Thirdly, the patient stays in the hospital for a certain time. The attribute *Stay* depends on all the other seven attributes. The dependency among the attributes in this fracture database, based on Wong and Leung's description of the data set, is shown in the directed graph of Figure 7.5. In the figure an attribute Y depends on an attribute X if and only if there is a path from X to Y in the directed graph.

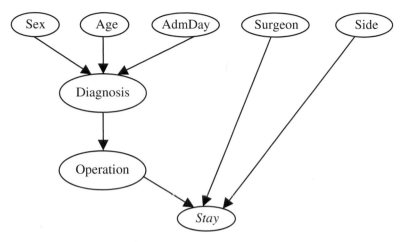

Figure 7.5: Network of dependency among attributes in a fracture database

A small part of the grammar for this data set, taking into account the network of attribute dependency shown in Figure 7.5, is shown in Figure 7.6. A more complete view of this grammar can be found in [Wong and Leung 2000]. In Figure 7.6 terminal symbols of the grammar are written in **bold** font style. The first three grammar rules in the figure are justified by the fact that the user wants rules predicting one of the following three goal attributes: *Diagnosis, Operation* and *Stay*. The next three grammar rules specify that each of the above three prediction rules consists of an antecedent (Antec) and a consequent (Cons). The next grammar rules specify the structure of each antecedent and each consequent.

Each antecedent is a conjunction of attribute-value descriptors, whereas each consequent is a single goal-attribute-value descriptor. Finally, each descriptor is associated with a grammar rule which specifies the attribute-value conditions that can be taken on by the descriptor, based on the values belonging to the domain of each attribute. The symbol "|" is used to indicate alternative definitions for a descriptor. Note that the definition of a descriptor of a rule antecedent will typically include a grammar rule that produces the keyword "**any**", which is a wild card ("don't care" symbol) meaning that any value of the corresponding attribute satisfies the rule condition in question. This effectively means that the attribute is removed from the rule antecedent.

The disadvantage of this approach is that the grammar is specific to a given application domain. It is possible, however, to define a more generic grammar, perhaps more accurately described as a grammar template, or language template. The template can then be instantiated with details of a specific problem domain or a specific data sct, to produce a grammar that can be effectively used in rule discovery. In principle this instantiation could be done by automatically accessing a metadata file, with descriptions (e.g., the domain) of attributes and other data set-dependent information.

```
start => DiagnosisRule
start => OperationRule
start => StayRule
DiagnosisRule => IF DiagnosisAntec THEN DiagnosisCons
OperationRule => IF DiagnosisAntec AND OperationAntec
                   THEN OperationCons
StayRule => IF DiagnosisAntec AND OperationAntec AND StayAntec
              THEN StayCons
DiagnosisAntec => SexDescriptor AND AgeDescriptor
                   AND AdmDayDescriptor
DiagnosisCons => DiagnosisDescriptor
  .
  . (similar definitions for OperationAntec,
    OperationCons, StayAntec, StayCons)
  .
SexDescriptor => "any"
SexDescriptor => "Sex = male" | "Sex = female"
  .
  . (similar definitions for other attribute descriptions)
  .
```

Figure 7.6: Part of the grammar for fracture data set described in Figure 7.5

One example of this kind of generic grammar template is shown in Figure 7.7, adapted from [Podgorelec and Kokol 2000]. In this example, COND denotes condition, ATTR denotes attribute, ATTR_VAL denotes attribute value, EXPR denotes arithmetic expression, CATEG denotes categorical, CONT denotes continuous, REL_OP denotes relational comparison operator, and OP denotes mathematical operator. Again, terminal symbols are shown in **bold**.

As an example of the generality of this grammar template, each non-terminal CATEG_ATTR or CONT_ATTR can be expanded to any categorical or continuous attribute of the data set being mined. Of course, this generality is obtained at the expense of losing the ability to represent background knowledge that is specific to a given application domain.

```
COND => CATEG_COND | CONT_COND
CATEG_COND => CATEG_ATTR  CATEG_REL_OP
                  CATEG_ATTR_VAL
CONT_COND => EXPR  CONT_REL_OP  EXPR
CATEG_REL_OP => = | ≠
CONT_REL_OP => < | ≤ | > | ≥
EXPR => EXPR OP EXPR | ( EXPR ) | CONT_ATTR |
            CONT_ATTR_VAL
OP => + | - | × | /
CATEG_ATTR => one of the possible categorical attributes
CONT_ATTR => one of the possible continuous attributes
CATEG_ATTR_VAL => one of the possible values
                  of a categorical attribute
CONT_ATTR_VAL => one of the possible values
                  of a continuous attribute
```

Figure 7.7: Example of a generic grammar template for rule discovery

7.5 GP for Decision-Tree Building

Assuming that GP individuals are represented by a tree, as usual in the vast majority of the literature, intuitively GP can be used to build decision trees in a more natural way than other kinds of evolutionary algorithms. Of course, this does not mean that the best use of GP for data mining applications consists of building decision trees. GP can also be applied to the discovery of rules not expressed in a decision-tree form, as discussed in the previous sections. In addition, any user considering the use of GP for rule discovery should carefully consider the pros and cons of decision trees discussed in subsection 3.1.3.

The basic ideas involved in using GP for building a decision tree are as follows [Koza 1992, chap. 17]. The terminal set consists of the names of the classes (goal-attribute values) to be predicted. The function set consists of the names of predictor attributes. Actually, each predictor attribute name is considered (by GP) as a "function", whose returned value depends on the value of the corresponding attribute in the data being mined. More precisely, each internal (non-leaf) decision-tree node is associated with a predictor-attribute name and a test over the values of that attribute, as discussed in section 3.1.1. Therefore, each attribute-name "function" has as many arguments as the number of possible outcomes of the corresponding test.

The basic idea of using GP for evolving decision trees has been present in several recent projects. For instance, both [Ryan and Rayward-Smith 1998] and [Fu 1999] propose a kind of hybrid GA/GP system, where each individual of the

GA is a decision tree. In essence, both systems use the well-known decision-tree building algorithm C4.5 [Quinlan 1993] to generate decision trees that are used as individuals by the GA/GP system, though Ryan and Rayward-Smith's system uses a modified version of C4.5 that randomly selects attributes for the internal decision-tree nodes, rather than using a local heuristic for that selection. Both systems apply genetic operators such as crossover directly on the trees generated by C4.5. In addition, both systems use some special-purpose procedures to guarantee that each new tree produced by crossover or mutation is valid. Such procedures can be considered as GP editing operators, and they are well described and justified in [Ryan and Rayward-Smith 1998].

One important difference between the two systems is as follows. The system proposed by Fu was designed mainly for large-scale data mining applications, using data samples (rather than all available data) to build decision trees, and a decision tree's quality is evaluated with respect to predictive accuracy only. In contrast, the system proposed by Ryan and Rayward-Smith was designed mainly to build decision trees that are both accurate and small (and so hopefully more comprehensible for the user), by using a fitness function that considers both a tree's predictive accuracy and its size.

Another kind of hybrid GA/GP for evolving decision trees is proposed by [Papagelis and Kalles 2001]. This system also was designed to discover decision trees that are both accurate and small. Unlike the two previous systems, this system does not use C4.5 nor another off-the-shelf decision-tree-building algorithm. The search for decision trees is entirely performed by an evolutionary algorithm. Even the choice of the class to be assigned to a leaf node is made at random. Instead of this random choice of classes, perhaps it would be better to label a leaf node with the most frequent class among the data instances belonging to that node. This is a simple way of incorporating task-dependent knowledge into an evolutionary algorithm for classification, helping to maximize the fitness of an individual, as discussed in subsection 6.1.3.2.

In the next subsections we briefly discuss other GP systems for building decision trees that depart from the above-mentioned GPs in several different ways.

7.5.1 Evolving Decision Trees with First-Order-Logic Conditions

Another GP for evolving decision trees was proposed by [Rouwhorst and Engelbrecht 2000]. Unlike the above-discussed GPs for evolving decision trees, this GP builds a decision tree where an internal node can contain a first-order-logic condition (section 1.2.1). More precisely, an internal node contains either a propositional condition in the form $(Attr_i\ Op\ Val_{ij})$ or a first-order-logic condition in the form $(Attr_i\ Op\ Attr_j)$, where $Attr_i$ and $Attr_j$ are the i-th and j-th predictor attributes, Op is some comparison operator, and Val_{ij} is the j-th value of $Attr_i$'s domain.

The use of first-order-logic conditions increases the expressiveness power of the decision tree. This seems one of the main reasons why the decision trees built by this GP were considerably smaller – while still being competitive in terms of predictive accuracy – than the rule set and decision tree built by other two algorithms (CN2 and C4.5).

Apparently, however, this GP generates first-order-logic conditions without respecting elaborate data type restrictions. For instance, one of the examples of first-order-logic condition mentioned in [Rouwhorst and Engelbrecht 2000, p. 634] is (*Age* ≥ *Bloodsugar_level*). Although both *Age* and *Bloodsugar_level* are continuous attributes, comparing values of these attributes with a "≥" operator is not meaningful for the user, since these two attributes have very different, incompatible attribute domains.

When working with first-order-logic conditions, one approach to ensure that only meaningful conditions (with respect to data type combinations) are generated is to specify an Attribute-Compatibility Table, as mentioned at the end of section 7.3.

7.5.2 Evolving Linear Decision Trees

As discussed in subsection 3.1.5, in a linear decision tree the tests on attribute values performed at each tree node are linear combinations of some of the (numeric) attributes. Therefore, the nodes represent oblique cuts in data space, in contrast with the axis-parallel cuts produced by conventional decision tree algorithms. Recall that this makes oblique decision tree algorithms more flexible, at the expense of reducing the comprehensibility of the discovered rules, in comparison with axis-parallel decision trees. This loss of comprehensibility is probably the main reason why oblique decision trees are much less popular than axis-parallel ones.

In any case, the search space of oblique decision tree algorithms is much larger than the search space of axis-parallel decision tree algorithms. This can be considered a motivation for using an evolutionary algorithm to build an oblique decision tree, assuming that evolutionary algorithms tend to be more effective than local search algorithms in very large search spaces.

[Bot and Landgon 2000] proposed strong typing GP for evolving linear decision trees. As discussed in section 7.3, the basic idea of strong typing is that, for each function in the function set, one specifies both the data type of each argument and the data type of the value returned by the function. In addition, one specifies the data type for each terminal in the terminal set. Whenever an individual is generated – e.g., when generating the initial population, or when crossover and/or mutation is performed – the data type restrictions are enforced, which guarantees that the generated individual is syntactically valid.

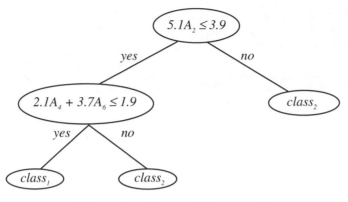

(a) linear decision tree ("phenotype")

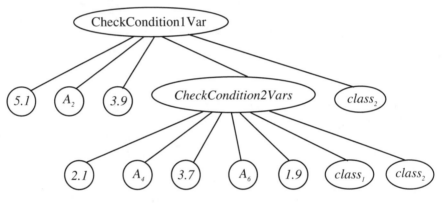

(b) genetic representation of the tree shown in (a)

Figure 7.8: GP individual representing a linear decision tree

In Bot and Langdon's GP the function set includes 3 kinds of functions: CheckCondition1Var, CheckCondition2Vars, and CheckConditions3Vars, with their obvious meaning of checking conditions with 1, 2 and 3 variables (attributes), respectively. These functions have 5, 7, and 9 arguments, respectively. Let k be the number of "conditions being checked", i.e., the number of attributes being combined. The first $2 \times k$ arguments of a function, $k = 1,2,3$, represent an attribute index and its respective coefficient in the linear combination. The $((2 \times k)+1)$-th argument of a function represents the threshold (a constant) of the inequality. The last two arguments of a function contain either a predicted class or another function, which is recursively called to further expand the decision tree. For instance, in Figure 7.8(a) the internal nodes $(5.1A_2 \leq 3.9)$ and $(2.1A_4 + 3.7A_6 \leq 1.9)$ are functions of type CheckCondition1Var and CheckCondition2Vars, respectively, and are represented in a GP individual as shown in Figure 7.8(b).

It is also possible to use other kinds of EAs (other than GP) to help building a linear (or oblique) decision tree. For instance, [Cantu-Paz and Kamath 2000] use a simple EA – a (1+1) evolution strategy – and a GA to optimize the hyperplanes represented by oblique decision tree nodes. The basic idea is that a candidate hyperplane is represented as a vector of real-valued coefficients, and these coefficients are optimized by the (1+1) evolution strategy or the GA.

7.5.3 Hybrid GP, Cellular Automaton, and Simulated Annealing

[Folino et al. 2000] proposed to induce a decision tree using a very hybrid method, combining GP, cellular automaton and simulated annealing. In essence, each cell of the cellular automaton (CA) corresponds to a GP individual, which is a decision tree. Each individual interacts only with its neighbors in a two-dimensional CA grid. In their experiments the authors use a 20x20 grid, corresponding to a GP population of 400 individuals/decision trees. Information contained in GP individuals slowly diffuses across the grid, so that clusters of solutions can be formed around different optima in the fitness landscape.

When an individual undergoes crossover, a mate is selected as the neighbor individual having the best fitness. The current individual is then replaced or not by the offspring having the best fitness depending on whether or not the fitness increase associated with this replacement is less than or equal to the current "temperature" of the system. This temperature is a function of the percentage of data instances correctly classified. The temperature is gradually reduced, as usual in simulated annealing, and the use of this mechanism is supposed to avoid the problem of premature convergence of GP.

The authors argue that this approach has a good potential for implementation on a massively-parallel processing system, thus being suitable for large-scale data mining. Though the experiments reported in the paper involve a sequential simulation of the CA and small data sets (whose number of data instances varies from 150 to 3772), the authors mention that a parallel implementation of the method is in progress.

7.5.4 Using GP for Implementing Decision-Tree Nodes

The previous subsections discussed GP algorithms for evolving decision trees. In general those GPs work by associating one individual with an entire decision tree. A different approach consists of incrementally building a decision tree by using a GP to implement the test (decision) performed at each internal decision-tree node [Marmelstein and Lamont 1998].

In this approach, starting with the root node of the decision tree, at each tree node a GP is run to find a test over attribute values. As usual in decision trees, the

GP must find a test whose outcomes discriminate the classes of the data instances belonging to the current tree node.

In comparison with a standard GP for classification (subsection 5.4.4), this system has the advantage of producing trees that tend to be somewhat more comprehensible for the user. After all, each solution found by a GP (in each internal decision-tree node) can be separately analyzed, and tends to be less complex than the entire tree produced by a standard GP for classification. On the other hand, apparently the system produces decision trees that are more complex than conventional decision trees. To quote [Marmelstein and Lamont 1998, p. 230]:

"There is ... no guarantee that the decision boundaries produced will be readily understandable. Interpreting the partitions formed by even simple GPs can be a cumbersome task, requiring the assistance of a domain expert."

7.6 On the Quality of Rules Discovered by GP

In general GP can be naturally used to discover prediction rules whose knowledge representation has great expressiveness power, by comparison with the rule representation used by most other kinds of evolutionary algorithms.

For instance, in most GAs for rule discovery the set of operators encoded into the genotype of an individual is limited to comparison operators in the set $\{=, \neq, <, \leq, >, \geq\}$, and the encoding of first-order-logic rule conditions (subsection 1.2.1) into a conventional GA individual can be somewhat cumbersome.

In contrast, in GP the function set can easily contain functions/operators of arbitrary complexity, and the handling of first-order-logic rule conditions tends to be somewhat less cumbersome than in conventional GAs. It should be pointed out, however, that the greater flexibility offered by GP, allowing the user to include an arbitrarily-complex function in the function set, has been somewhat under-explored in most applications of GP for rule discovery.

Another potential advantage of GP for rule discovery has to do with the degree of interestingness of the discovered rules. Recall that, as discussed in subsection 1.1.1 and section 2.3, in data mining we usually want to discover rules with a high degree of: (a) accuracy; (b) comprehensibility (or simplicity); and (c) interestingness. One aspect of rule interestingness involves the degree of surprisingness, or novelty, for the user. In other words, ideally a rule should represent a data relationship which was previously unknown to the user.

GP has the potential to produce rules that are truly surprising and interesting for the user – while at the same time having a good predictive accuracy. This seems at least partially due to GP's ability to discover important attribute interactions, as a result of applying several different functions to the original attributes. It is quite possible that, although one attribute by itself is little relevant for prediction, that attribute can be actually relevant for prediction when combined, via some function, with other attributes. A good example of GP's ability to discover surprising (yet accurate) knowledge can be found in [Gilbert et al. 1998, pp. 113-114]:

"Surprisingly, the GP frequently selects variables other than those indicated by the statistical metrics to be the most characteristic for classification purposes ... In fact, in addition to those that are most characteristic, the GP consistently selects variables that are least *characteristic." (p.113)*

"The GP was able to derive rules capable of modeling the data using just 6-9 variables ... Inspection of the GP 'derived rules showed that the models used not only the most characteristic variables (as measured by standard statistical metrics) but also the least *characteristic. This enabled the GPs to derive good predictive models... The GP was also able to identify variables which are actually more characteristic for classification purposes than the statistical metrics would suggest." (p. 114)*

It is interesting to note that the rules discovered by GP can be novel, interesting and useful for the user even though they do not have a high predictive accuracy. For instance, Wong and Leung applied a grammar-based GP (section 7.4) to a real-world scoliosis data set, using a certain rule interestingness measure to support the discovery of interesting rules. Some interesting results obtained by the authors are as follows [Wong and Leung 2000, p. 166]:

"...the system found rules with confidence factors [a measure of predictive accuracy] around 40% to 60%. Nevertheless, the rules...show something different in comparison with the rules suggested by the clinicians [physicians]...After discussing with the domain expert, it is agreed that the existing rules are not defined clearly enough, and our rules are more accurate than theirs. Our rules provide hints to the clinicians to re-formulate their concepts."

On the other hand, the use of GP for rule discovery also has some potential disadvantages. Some researchers have noted that GP has a low generalization ability in some cases, i.e., the error in the prediction of data instances of the test set, unseen during training, is considerably higher than the error in the prediction of training data instances. [Keijzer and Babovic 2000, p. 81] go so far as to state that:

"Genetic programming often produces models with extremely poor generalization properties. ... When the resulting formula of a single rule run is applied to unseen data, if often happens that it is undefined at certain points or has extreme values..."

The authors present some evidence that this *overfitting* problem can occur in regression problems, particularly when the function set is unable to perfectly represent the true regression function, i.e., when the function set does not satisfy the sufficiency property. Although such cases are not often discussed in the literature, they are quite possible in practice. After all, since arguably the main goal of data mining is essentially to search for *unknown* relationships, one cannot always assume that the function set satisfies the sufficiency property, at least not to a large degree. Overfitting problems can also be caused by other factors, as discussed in subsection 2.1.2.

To cope with the overfitting problem, Keijzer and Babovic propose an ensemble method, consisting of running GP several times, each time with a different training set sample, and then combining the results of all runs into a single average prediction model. In addition, in order to produce the average model, the authors recommend to use a "trimming" procedure to avoid that the extreme predictions of some runs dominate the predictions made by the average model. More precisely, they suggest that the highest and lowest predictions with a pre-specified percentage, say 10%, be excluded from the computation of the average prediction.

In any case, the fact that GP produces rules (or models) with low generalization ability in some cases can hardly be considered an inherent disadvantage of GP, in the sense that any prediction-rule discovery algorithm will produce rules with low generalization ability in some cases. This is a direct consequence of the fact that any prediction-rule discovery algorithm has an inductive bias, and any inductive bias is suitable for some data sets and unsuitable for other data sets, as discussed in section 2.5.

On the other hand, the tendency of GP for producing large rule sets (since the size of GP trees often grow a lot during evolution) can be considered a disadvantage from the viewpoint of rule simplicity (or comprehensibility). Actually, using GP to discover simple classification rules can be considered a research issue. One approach to cope with this issue is briefly discussed next.

In the following we assume that rule simplicity is inversely proportional to rule length (number of rule conditions). Rule length is by far the most used criterion to evaluate rule simplicity in the literature. However, one must bear in mind that a precise evaluation of rule simplicity is a very difficult problem, involving some subjective issues [Pazzani 2000]. Measuring rule length is just a simple, objective (but imperfect) way of estimating rule simplicity.

An often-used approach to favor the discovery of short rules is to include a penalty for long rules and/or many rules in the fitness function. An example of the use of this approach can be found in [Bojarczuk et al. 2000], where a part of the fitness function is a direct measure of rule simplicity, given by the formula:

$$Simplicity = (max\_nodes - 0.5 num\_nodes - 0.5) / (max\_nodes - 1),$$

where max_nodes is the maximum allowed number of nodes of a tree (individual) and num_nodes is the current number of nodes of the tree. This formula produces its maximum value of 1.0 when a rule is so simple that it contains just one term, and it produces its minimum value of 0.5 when the number of tree nodes equals the allowed maximum. This formula is multiplied to another formula that measures the predictive accuracy of the rules encoded in the individual, and the higher the value of this multiplication the higher the fitness of the individual. Since in this system the goal is to maximize fitness, the above formula has the role of penalizing individuals containing long rules and/or many rules.

Conceptually similar penalty terms are also used by other GP systems for rule discovery, even though the used formula is quite different. The basic idea is that the larger the number of nodes in the individual, the larger the penalty. For example, in [Gilbert et al. 1998] the penalty is given by the formula: 0.01 x *num _nodes*. As another example, [Bot and Langdon 2000] foster rule simplicity by lowering the fitness value by some factor times the number of nodes in the tree or its depth. They compared two approaches for this kind of penalization, namely using a penalty based on the number of tree nodes or on the tree depth. They report no statistical difference in classification accuracy when either approach is used.

In any case, one should bear in mind that using a fitness function that favors the discovery of shorter rules is a form of inductive bias (section 2.5), and therefore, with respect to the maximization of predictive accuracy, the effectiveness of this approach is strongly dependent on the application domain. For instance, [Cavaretta and Chellapilla 1999] report computational results indicating that, in one credit data set, a no-complexity-bias GP produced more complex and more accurate models (trees) than a low-complexity-bias GP – a GP penalizing larger trees.

Finally, it should be noted that the power of GP for data mining has been under-explored in the literature. In particular, each GP run (like a GA run) is normally used to discover a rule set for a given data set. The discovered rule set can be called a "program" only in a loose sense. A more interesting (and much more difficult) usage of GP would be to discover a generic data mining algorithm, which could then be used for actually mining different data sets from different application domains. This would probably require that a single GP run accesses many different data sets from different application domains, in order to produce a truly generic data mining algorithm. In this case one would be really performing program (algorithm) induction – the discovered algorithm would be a general "recipe" for mining data sets, rather than a single rule set induced for one particular data set. We elaborate a little more this idea in subsection 12.2.3.

References

[Banzhaf et al. 1998] W. Banzhaf, P. Nordin, R.E. Keller, and F.D. Francone. *Genetic Programming ~ an Introduction: On the Automatic Evolution of Computer Programs and Its Applications.* Morgan Kaufmann, 1998.

[Bhattacharyya et al. 1998] S. Bhatacharyya, O. Pictet and G. Zumbach. Representational semantics for genetic programming based learning in high-frequency financial data. *Genetic Programming 1998: Proceedings of the 3rd Annual Conference (GP '98)*, 11–16. Morgan Kaufmann, 1998.

[Bojarczuk et al. 1999] C.C. Bojarczuk, H.S. Lopes, and A.A. Freitas. Discovering comprehensible classification rules using genetic programming: a case study in a medical domain. *Proceedings of the Genetic and Evolutionary Computation Conference (GECCO '99)*, 953–958. Morgan Kaufmann, 1999.

[Bojarczuk et al. 2000] C.C. Bojarczuk, H.S. Lopes, and A.A. Freitas. Genetic programming for knowledge discovery in chest pain diagnosis. *IEEE Engi-*

neering in Medicine and Biology Magazine, 19(4) – special issue on data mining and knowledge discovery, 38–44, July/Aug. 2000.

[Bojarczuk et al. 2001] C.C. Bojarczuk, H.S. Lopes and A.A. Freitas. Data mining with constrained-syntax genetic programming: applications in medical data sets. *Proceedings of the Intelligent Data Analysis in Medicine and Pharmacology – a workshop at MedInfo-2001*. London, UK, 2001.

[Bot and Langdon 2000] M.C.J. Bot and W.B. Langdon. Application of genetic programming to induction of linear classification trees. *Genetic Programming: Proceedings of the 3rd European Conference (EuroGP 2000). Lecture Notes in Computer Science 1802*, 247–258. Springer, 2000.

[Cantu-Paz and Kamath 2000] E. Cantu-Paz and C. Kamath. Using evolutionary algorithms to induce oblique decision trees. *Proceedings of the Genetic and Evolutionary Computation Conference (GECCO '2000)*, 1053–1060. Morgan Kaufmann, 2000.

[Cavaretta and Chellapilla 1999] M.J. Cavaretta and K. Chellapilla. Data mining using genetic programming: the implications of parsimony on generalization error. *Proceedings of the 1999 Congress on Evolutionary Computation (CEC '99)*, 1330–1337. IEEE Press, 1999.

[Eggermont et al. 1999a] J. Eggermont, A.E. Eiben and J.I. van Hemert. Adapting the fitness function in GP for data mining. *Proceedings of the 1999 European Conference on Genetic Programming (EuroGP '99)*. 1999.

[Eggermont et al. 1999b] J. Eggermont, A.E. Eiben and J.I. van Hemert. A comparison of genetic programming variants for data classification. *Proceedings of the 1999 Conference on Intelligent Data Analysis (IDA '99)*.

[Folino et al. 2000] G. Folino, C. Pizzyti, and G. Spezzano. Genetic programming and simulated annealing: a hybrid method to evolve decision trees. *Genetic Programming: Proceedings of the 3rd European Conference (EuroGP 2000). Lecture Notes in Computer Science 1802*, 294–303. Springer, 2000.

[Freitas 1998] A.A. Freitas. A genetic algorithm for generalized rule induction. In: R. Roy et al. *Advances in Soft Computing – Engineering Design and Manufacturing. (Proceedings of the WSC3, 3rd on-line world Conference, hosted on the internet, 1998)*, 340–353. Springer, 1999.

[Fu 1999] Z. Fu. An innovative GA-based decision tree classifier in large scale data mining. *Proceedings of the 3rd European Conference on Principles and Practice of Knowledge Discovery in Databases (PKDD '99)*, 348–353. 1999.

[Gilbert et al. 1998] R.G. Gilbert, R. Goodacre, B. Shann, D.B. Kell, J. Taylor and J.J. Rowland. Genetic programming-based variable selection for high-dimensional data. *Genetic Programming 1998: Proceedings of the 3rd Annual Conference (GP '98)*, 109–115. Morgan Kaufmann, 1998.

[Hu 1998] Y.-J. Hu. A genetic programming approach to constructive induction. *Genetic Programming 1998: Proceedings of the 3rd Annual Conference (GP '98)*, 146–151. Morgan Kaufmann, 1998.

[Keijzer and Babovic 1999] M. Keijzer and V. Babovic. Dimensionally aware genetic programming. *Proceedings of the Genetic and Evolutionary Computation Conference (GECCO '99)*, 1069–1076. Morgan Kaufmann, 1999.

[Keijzer and Babovic 2000] M. Keijzer and V. Babovic. Genetic programming, ensemble methods and the bias/variance tradeoff – introductory investigations. *Genetic Programming: Proceedings of the 3rd European Conference (EuroGP 2000). Lecture Notes in Computer Science 1802*, 76–90. Springer, 2000.

[Koza 1992] J.R. Koza. *Genetic Programming: on the programming of computers by means of natural selection.* MIT Press, 1992.

[Marmelstein and Lamont 1998] R.E. Marmelstein and G.B. Lamont. Pattern classification using a hybrid genetic program – decision tree approach. *Genetic Programming 1998: Proceedings of the 3rd Annual Conference (GP '98)*, 223–231. Morgan Kaufmann, 1998.

[Mendes et al. 2001] R.F. Mendes, F.B. Voznika, A.A. Freitas and J.C. Nievola. Discovering fuzzy classification rules with genetic programming and co-evolution. *Principles of Data Mining and Knowledge Discovery (Proceedings of the 5th European Conference, PKDD 2001) – Lecture Notes in Artificial Intelligence 2168*, 314–325. Springer, '2001.

[Montana 1995] D.J. Montana. Strongly Typed Genetic Programming. *Evolutionary Computation 3(2)*, 199–230. 1995.

[Papagelis and Kalles 2001] A. Papagelis and D. Kalles. Breeding decision trees using evolutionary techniques. *Proceedings of the 18th International Conference on Machine Learning (ICML '2001)*, 393–400. Morgan Kaufmann, 2001.

[Pazzani 2000] M.J. Pazzani. Knowledge discovery from data? *IEEE Intelligent Systems 15(2)*, 10–12, Mar./Apr. 2000, .

[Podgorelec and Kokol 2000] V. Podgorelec and P. Kokol. Fighting program bloat with the fractal complexity measure. *Genetic Programming: Proceedings of the 3rd European Conference (EuroGP 2000). Lecture Notes in Computer Science 1802*, 326–337. Springer, 2000.

[Quinlan 1993] J.R. Quinlan. *C4.5: Programs for Machine Learning.* Morgan Kaufmann, 1993.

[Rouwhorst and Engelbrecht 2000] S.E.Rouwhorst and A.P. Engelbrecht. Searching the forest: using decision trees as building blocks for evolutionary search in classification databases. *Proceedings of the 2000 Congress on Evolutionary Computation (CEC '2000).* IEEE, 2000.

[Ryan and Rayward-Smith 98] M.D. Ryan and V.J. Rayward-Smith. The evolution of decision trees. *Genetic Programming 1998: Proceedings of the 3rd Annual Conference (GP '98)*, 350–358. Morgan Kaufmann, 1998.

[Ryu and Eick 1996] T.-W. Ryu and C.F. Eick. MASSON: discovering commonalities in collection of objects using genetic programming. *Genetic Programming 1996: Proceedings of the 1st Annual Conference (GP '96)*, 200–208. MIT Press, 1996.

[Wong and Leung 2000] M.L. Wong and K.S. Leung. *Data Mining Using Grammar Based Genetic Programming and Applications.* Kluwer, 2000.

8 Evolutionary Algorithms for Clustering

> "The act of sorting similar things into categories is one
> of the most primitive and common pursuits of men."
> Jain and Dubes [Backer 1995, p. 3]

In section 2.4 we reviewed the basic ideas of two major types of clustering methods, namely iterative-partitioning and hierarchical methods. In this chapter we discuss several issues in the development of Evolutionary Algorithms (EAs) for clustering following the iterative-partitioning approach, which seems the most common approach in the EA literature. (In passing, we mention that a couple of recent projects on GAs for clustering following the hierarchical approach can be found in [Rizzi 1998; Lozano and Larranaga 1999].)

In order to facilitate our study of EAs for clustering, we have divided them into three broad groups based on the kind of individual representation used by the EA, since this representation determines the kind of solution returned by the EA. More precisely, this chapter is divided into five sections, as follows. Section 8.1 discusses EAs where an individual represents cluster descriptions, specifying, e.g., the shape and size of clusters. Section 8.2 discusses EAs where an individual represents the coordinates of clusters' centroids or medoids. Section 8.3 discusses EAs with an instance-based individual representation. Section 8.4 discusses fitness evaluation. Finally, section 8.5 briefly discusses the pros and cons of EAs for clustering, by comparing them with conventional, local-heuristic clustering techniques.

8.1 Cluster Description-Based Individual Representation

In this section we discuss EAs for clustering where an individual represents a set of clusters descriptions. That is, an individual contains values of parameters that describe the shape, size and possibly other aspects of a set of clusters.

We emphasize that in clustering we normally want to find a *set* of clusters, rather than a single cluster – assuming, of course, that there are two or more distinct groups of data to be found in the data set being mined. At a very high level of abstraction, this is similar to the desire for discovering a *set* of prediction rules – each rule covering a different part of the entire data space – in the classification task.

Hence, in clustering we face a similar choice between the Pittsburgh and Michigan approaches that we discussed in chapter 6. We can associate an EA individual with a set of clusters, following the Pittsburgh approach, or we can associate an EA individual with a single cluster and have a set of individuals (possibly the entire population) represent a set of clusters. The pros and cons of each approach are conceptually similar to the ones discussed in chapter 6 in the context of classification.

We discuss in the next two subsections EAs following the Pittsburgh approach, which has been more common in the literature. More precisely, in subsection 8.1.1 we discuss the issue of individual encoding for representing cluster descriptions, whereas in subsection 8.1.2 we discuss the issue of genetic operators suitable for this kind of individual representation.

8.1.1 Individual Representation

There are many different shapes of clusters, such as spheres, ellipsoids, boxes, etc. In principle a GA for finding cluster descriptions could search for clusters of different shapes, but the size of the search space would be very large. As a result, typically the GA searches for cluster descriptions of a single type. We review below two approaches, one based on ellipsoids and the other one based on boxes.

[Srikanth et al. 1995] proposed a Pittsburgh-style GA for clustering where each individual contains a set of ellipsoid-shaped cluster descriptions. An individual has a variable length, since the number of clusters encoded into an individual is variable. Alternatively, one could use a fixed-length string if the user could specify the desired number of clusters in advance.

In Srikanth et al.'s work each cluster description consists of a set of parameters specifying the size and shape of an ellipsoid. More precisely, an ellipsoid is completely specified by the following kinds of parameters:

- the coordinates of the ellipsoid's origin (or center);
- the length of the ellipsoid's axes; and
- the ellipsoid's orientation (rotation angle) with respect to the axes of reference.

In this system the above parameters are encoded into an individual of a GA by using a binary encoding, where each parameter is encoded by a certain number of bits. Alternatively, one could use a real-valued encoding, where each parameter is directly encoded by a real-valued number. In any case we can think of an ellipsoid as encoded into a fixed-length (sub)string, and an individual consists of a sequence of ellipsoid descriptions.

[Ghozeil and Fogel 1996] proposed a Pittsburgh-style evolutionary programming algorithm for clustering where each individual contains a set of hyperbox-shaped cluster descriptions. Each cluster description was represented by five parameters, namely (x, y, θ, w, h), where:

- x and y define the (x,y) position of the center of the hyperbox;
- θ is the rotation (in radians) anti-clockwise of the hyperbox around its center;
- w is the length of the sides of the hyperbox in the direction of θ (width) and h is the length of the sides of the hyperbox in the direction perpendicular to θ (height).

Note that these parameters are conceptually similar to the ones used by Srikanth et al. to specify ellipsoid descriptions. Generalizing from both projects, Figure 8.1 shows, at a high level of abstraction, the structure of an individual specifying a set of cluster descriptions, following the Pittsburgh approach.

description of cluster 1   description of cluster K

center position	orientation (angle)	length of axes / sides		center position	orientation (angle)	length of axes / sides
			. . .			

Figure 8.1: High-level structure of an individual specifying cluster descriptions

One difference between the above two projects is that in Ghozeil and Fogel's work individuals contained not only cluster description parameters but also other kinds of parameters. More precisely, each individual also contained five self-adaptive parameters, each of them a standard deviation used in the perturbation (mutation) of the above-mentioned five parameters that specify a hyperbox. Each parent produced a single offspring. For each hyperbox in a parent individual, a corresponding hyperbox for the offspring was created by adding a Gaussian random variable to the parent's hyperbox parameters.

In addition, each individual used a parameter NBox to define the number of hyperboxes – the number of clusters – in its candidate solution. The NBox parameter is subject to mutation and evolution, so that this system uses a variable-length genotype. Individuals also contained two self-adaptive mutation factors controlling how often a box is added or deleted in the creation of an offspring.

8.1.2 Genetic Operators

8.1.2.1 Adapting Crossover for the Pittsburgh Approach

Once parameters specifying the size and shape of clusters are encoded into an individual's genotype, there are several crossover operators proposed in the literature that could be applied to an individual, with little or no modification. For instance, [Srikanth et al. 1995] use an one-point crossover operator adapted to the Pittsburgh approach whose basic idea is shown in Figure 8.2.

In this figure the crossover points are denoted by the symbol "|". Each individual consists of a sequence of cluster descriptions, where each cluster description is specified by ten bits.

Note that although the crossover points can fall in different cluster descriptions in the two individuals, they are restricted to fall on the same position within each cluster description. For instance, in Figure 8.2 the crossover points fell in the first and the second cluster descriptions of the two parent individuals, respectively, but both crossover points fell right after the third bit within each cluster description.

```
parent 1:  1 0 0 | 1 1 1 1 1 0 0    1 1 0  1 0 1 0 1 1 0
parent 2:  0 0 1 ' 0 0 1 0 1 1 0    1 1 0 | 0 0 0 0 0 1 1    0 0 1 0 0 1 0 1 0 1

child 1:   1 0 0  0 0 0 0 0 1 1     0 0 1 0 0 1 0 1 0 1
child 2:   0 0 1  0 0 1 0 1 1 0     1 1 0 1 1 1 1 1 0 0    1 1 0 1 0 1 0 1 1 0
```

Figure 8.2: Example of crossover in a Pittsburgh-approach GA for clustering

8.1.2.2 Cluster-Insertion/Deletion Operator

In addition to adapting crossover by enforcing that crossover points in both parents are semantically equivalent, as discussed in the previous subsection, [Srikanth et al. 1995] also use cluster-insertion and cluster-deletion operators. (Strictly speaking, cluster-description-insertion and cluster-description-deletion operators.) Both operators are aware that an individual's genotype consists of a set of cluster descriptions, and manipulate entire cluster descriptions, i.e., they never cut a cluster description into smaller fragments.

The insert operator randomly selects a cluster description, or a sequence of contiguous cluster descriptions, in the genotype of an individual, called the donor. Then it inserts the donor's selected (sequence of) cluster descriptions into an insert-location on another individual, called the receiver. This insert-location is randomly chosen, subject to the constraint that it falls in between cluster descriptions. This process is illustrated in Figure 8.3, where a sequence of two clusters descriptions in the donor, shown between square brackets ("[ ]") in the figure, is inserted between the first and second cluster descriptions of a receiver. In this figure each group of six bits corresponds to a cluster description. Therefore, in this example the algorithm inserts two cluster descriptions of the donor into the receiver.

In addition, the system includes a deletion operator, which has the opposite effect of the insertion operator, i.e., it randomly deletes a (sequence of) cluster descriptions of an individual.

```
donor:      011010   001011 [111000  100110]

receiver:   100110   011100

            (a) before cluster insertion

donor:      011010    001011 111000    100110      (unaltered)

receiver:   100110   [111000  100110]   011100     (increased length)

            (b) after cluster insertion
```

Figure 8.3: Example of cluster-insertion operator in a GA for clustering

The above-discussed cluster-insertion/deletion operators directly increase/decrease the length of an individual, respectively. Hence, they could be used to control an individual's length, by making their probability of application dependent on the current length of the individual. More precisely, the probability of insertion (deletion) could be inversely proportional (directly proportional) to the individual's length. However, this possibility seems not to have been explored by the authors.

8.2 Centroid/Medoid-Based Individual Representation

In the above-discussed approaches for encoding cluster descriptions into an individual, a cluster description is in general specified by a set of parameters, which define the shape and size of a region representing a cluster in data space. For instance, the parameters might be a hypersphere's centroid coordinates and radius.

One alternative, somewhat implicit form of representing clusters is to specify, for each cluster, just its centroid or medoid coordinates. Then each data instance is assigned to the cluster represented by the centroid or medoid that is nearest to that instance.

In essence, the difference between the concepts of centroid and medoid is as follows. A centroid is simply a point whose coordinates are given by the arithmetic average of the corresponding coordinates of all the data instances belonging to the cluster. Note that the coordinates of the centroid usually are not the coordinates of any data instance of the data being mined. In contrast, a medoid is one of the original data instances available in the data being mined. A medoid is a representative data instance of its cluster, in the sense that it is the data instance which is nearest to the cluster's centroid.

The use of medoids tends to be more robust against outliers than the use of centroids [Krzanowski and Marriot 1995, p. 83]. Intuitively, this is analogous to the fact that medians are more robust against outliers than arithmetic means.

We discuss centroid-based and medoid-based individual representations separately in the next two subsections.

8.2.1 Centroid-Based Individual Representation

As mentioned above, one form of specifying a candidate solution for a clustering problem is to specify the coordinates, in data space, of the centroid of each cluster. In this case the goal of the clustering algorithm is to find an optimal set of centroids. There are several algorithms that can be used to solve this problem, including the well-known K-means algorithm, which was briefly reviewed in subsection 2.4.2.

In general this kind of algorithm has two drawbacks: it can be trapped by local maxima and can be sensitive to the (usually random) choice of initial candidate solution. A GA can be used to try to overcome these drawbacks, due to its global search nature.

The basic idea is to encode a set of centroids, each specified by a vector of coordinates in data space, into an individual's genotype. This basic idea allows the development of several clustering algorithms that combine the global search of a GA with the task-dependent heuristic of K-means.

One of the simplest approaches, proposed by [Babu and Murty 1993], is to have an individual of the GA correspond to the initial set of centroids given to the K-means algorithm. In this approach an individual is evaluated by simply running an entire K-means algorithm with the initial set of centroids encoded in its genotype. Hence, the GA is actually optimizing the selection of initial centroids – or "seed values" – for the K-means algorithm.

One problem of this approach is that it is very computationally expensive, since a full K-means algorithm must be run on the order of $P \times G$ times, where P is the population size and G is the number of generations of the GA.

A more elaborate approach, which involves running just one iteration of K-means for each individual, was recently proposed by [Hall et al. 1999]. In this algorithm an individual is represented by a matrix M with K rows and m columns, where K is the number of clusters and m is the number of attributes. The entry M_{ij} of the matrix – $i=1,...,K$ and $j=1,...,m$ – represents the value of the j-th coordinate of the i-th centroid. Once the centroid coordinates are specified, each data instance is assigned to the cluster represented by its nearest centroid.

The algorithm assumes all attributes are continuous, i.e., real-valued, so that centroid coordinates are also continuous. Centroid coordinates can be represented in an individual's genotype either in their original form of real value or in the form of a binary encoding. The authors report on both representations, but focus on the binary one. In particular, they achieved better results with a binary gray code representation, in which any two adjacent numbers are only one bit different [Deb 2000, 6-7], than with a conventional binary encoding.

In Hall et al.'s algorithm crossover is applied at the cluster level, as follows. The crossover operator is aware that each individual consists of a matrix with K rows and m columns, where K is the number of clusters and m is the number of attributes. For a given pair of parents, crossover is applied K times, once for each corresponding pair of clusters in the two parents. This is illustrated in Figure 8.4. As a result, the coordinates of all K centroids of the parents are affected by crossover, producing offspring that is potentially different from the parents with respect to all clusters.

The form of crossover used by Hall et al. is a two-point crossover acting on binary strings, but the basic scheme shown in Figure 8.4 is generic enough to be used with any other form of crossover – such as one-point crossover, uniform crossover, etc.

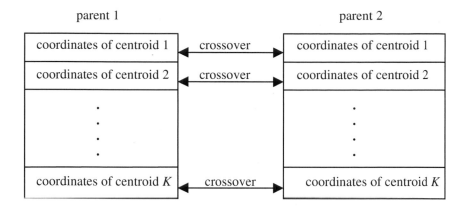

Figure 8.4: Crossover at cluster level for centroid-based individual encoding

8.2.2 Medoid-Based Individual Representation

The medoid-based approach simplifies the encoding of a candidate solution into an individual. After all, one can assign a unique sequential number in the range $1...n$ to each data instance, where n is the total number of instances in the data being mined, so that each medoid is completely specified by its unique sequential number.

A straightforward way of encoding a candidate solution – a set of medoids – into an individual's genotype would be to have a genotype with n binary genes, each of them taking on the value 1 or 0, indicating whether or not the respective data instance is chosen to be a cluster medoid. One problem with this approach, of course, is that it is not scalable with respect to the size (number of instances) of the data being mined, i.e., it is not practical when n is very large.

A more efficient, scalable approach is to encode a set of medoids as a string of integer-valued genes, where each gene is a distinct number in the range $1...n$. This approach is used, e.g., by [Estivill-Castro and Murray 1997], where an individual consists of a fixed-length string with K genes, where K is the number of clusters. Alternatively, this approach could be used with a variable-length string, so that the genetic search would consider solutions with a variable number of clusters.

In any case, a simple application of a conventional crossover operator can produce an invalid individual, having one or more repeated medoids. The authors avoid the generation of such invalid solutions by using a special crossover operator that tends to preserve common genes of parents – medoid numbers that occur in both parents – in the produced offspring.

8.3 Instance-Based Individual Representation

Assuming the number of clusters is predefined, a simple individual encoding for clustering consists of a genotype with n genes, where n is the number of data instances to be clustered. Each gene i – $i=1,...,n$ – contains an integer number j identifying the cluster to which the i-th instance is assigned – j is an integer number in the range $1,...,K$, where K is the number of clusters. This form of individual encoding is sometimes called a "string of group numbers" representation, but in this section we prefer the more generic term "instance-based" individual encoding.

It should be noted that this representation is not scalable for large data sets, since the genotype length is proportional to the number of instances to be clustered. Furthermore, this representation also has a high degree of redundancy and some problems associated with the application of conventional genetic operators, as discussed in [Falkenauer 1998, chap. 4]. As a very simple example of the high degree of redundancy in this kind of representation, consider the case where there are only four instances to be clustered and only two clusters, so that $n = 4$ and $K = 2$. The individuals (1 2 2 1) and (2 1 1 2) have very different genotypes. However, they represent the same clustering solution, i.e., a solution where the first and fourth instances are assigned to one cluster and the second and third instances are assigned to the other cluster. In addition, note that, although these two individuals represent the same clustering solution, applying a conventional crossover to these individuals can produce offspring representing clustering solutions quite different from these parents.

In any case, this representation has been used in some evolutionary algorithms for clustering [Meng et al. 2000; Krishma and Murty 1999]. In the next subsection we review the main aspects of one of these algorithms.

8.3.1 Hybrid Genetic Algorithm/*K*-Means

[Krishma and Murty 1999] have recently proposed a hybrid GA/K-means algorithm, called genetic K-means algorithm (GKA), following the above-mentioned instance-based individual encoding. The basic idea is to combine the global, population-based search of a GA with local, task-dependent operators.

In particular, instead of conventional crossover and mutation operators, GKA uses a K-means operator (KMO) and a distance-based mutation. KMO essentially consists of running one iteration of the K-means algorithm. In other words, given an individual's genotype – specifying which cluster each data instance is assigned to – KMO works in two steps. Firstly, it calculates the clusters' centroids based on the assignment of data instances to clusters specified by the individual. Second, it reassigns each data instance to the cluster whose centroid is nearest to that data instance.

Note that KMO is a deterministic operator. It has the advantage of incorporating task-specific knowledge into the GA. It has the disadvantage of performing a local search, so that its application can produce an individual that represents a local optimum in the space of all possible data partitions. Once a given individual reaches a local optimum it will not be changed by another application of KMO, so that there is no need to apply KMO on this individual again. However, the individual can escape from a local optimum if it is disturbed by the mutation operator, discussed below. As a result, KMO is useful mainly in the earlier generations of the GA. In later generations this operator becomes effective only when individuals are disturbed by mutation.

The distance-based mutation operator changes a gene value – the number of the cluster to which a data instance belongs – depending on the distances between that data instance and each of the clusters' centroids. Unlike KMO, this distance-based mutation operator is probabilistic. More precisely, the probability of changing a gene value to a given cluster number is inversely proportional to the distance between the data instance and that cluster's centroid. Therefore, this mutation operator favors the reassignment of a data instance to a cluster whose centroid is closer to that data instance, which is another way of incorporating task-specific knowledge into the GA.

8.3.2 Graph-Based Individual Representation

[Park and Song 1998] proposed an individual encoding that casts clustering as a graph partitioning problem. Although this is basically an instance-based individual encoding, it has some peculiarities that make it worthwhile to discuss it separately in this subsection. In essence, each data instance corresponds to a node in a graph, and a candidate solution consists of specifying connections among the graph nodes. Each subgraph, consisting of a subset of nodes that are connected to each other, corresponds to a cluster.

To encode a candidate solution Park and Song use an instance-based individual encoding that they call locus-based adjacency representation. An individual is represented by a vector of n integer elements, where n is the number of instances to be clustered. The i-th gene with value j means that nodes i and j are connected by a link in the graph. Hence, decoding an individual produces several subgraphs. As a very simple example, consider the following individual: $< 2, 1, 5, 3, 3 >$. Decoding this individual produces two subgraphs, and so two clusters, as shown in Figure 8.5.

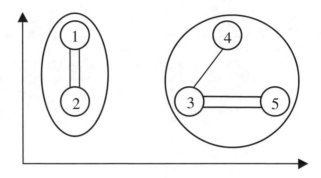

Figure 8.5: Graph-Based Clustering

There is a caveat in this representation. If each gene i – $i=1...n$ – can take on any value j – $j=1...n$, $j \neq i$ – then the size of the search space would be very large, namely $(n-1)^n$. Furthermore, in this case the search space would contain many candidate solutions that, although syntactically valid, do not make much sense. For instance, in the graph of Figure 8.5 it would not make sense to include nodes 1 and 5 in the same cluster, since they are very far from each other in the data space. Clearly, a good solution consists of clusters (subgraphs) where only data instances that are close to each other in the data space are grouped into the same cluster. This remark can be considered a kind of task-specific knowledge, which can be incorporated into a GA in order to make it somewhat "aware" of what constitutes a good solution for a clustering problem.

The authors have incorporated this kind of task-specific knowledge into the GA's individual representation. More precisely, each gene i can take on a value j only if the data instance j is one of the k nearest neighbors of data instance i, as measured by a given distance metric – e.g., Euclidean distance – in the data space. With this constraint, the size of the search space is greatly reduced to k^n, and now the search space contains only "sensible" candidate solutions.

Despite this significant improvement, it should be noted that this form of individual encoding still has one drawback inherent to any kind of instance-based individual encoding, namely the fact that it is not scalable for large data sets. As mentioned earlier, this individual encoding reduces scalability because the length of an individual is proportional to n, the number of instances to be clustered.

8.4 Fitness Evaluation

As discussed in section 2.4, evaluating the output of a clustering algorithm is a difficult task. This is partially due to the fact that, to some extent, the quality of a clustering solution is subjective. In practice most EAs for clustering ignore the subjective aspects involved in evaluating a clustering solution, and compute only an objective measure of clustering-solution quality.

In general, the first step is to decode the candidate clustering solution represented by an individual, in such a way that a partitioning matrix M is computed. This is a two-dimensional, $n \times K$ matrix, where n is the number of data instances in the data set being clustered and K is the number of clusters. Each cell M_{ij} of this matrix can take on either the value 1 or the value 0, indicating, respectively, whether or not the i-th data instance, $i=1,...,n$, is assigned to the j-th cluster, $j=1,...,K$.

In this section we assume that one desires clustering solutions with no overlapping clusters, i.e., where each data instance belongs to exactly one cluster, since this seems to be the most common case in practice. In this case a candidate clustering solution must satisfy the following constraint:

$$\sum_{j=1}^{K} M_{ij} = 1, \text{ for every } i\text{-th instance, } i=1,...,n.$$

The method used for decoding an individual in order to produce a partitioning matrix M obviously depends on the kind of individual representation used. The simplest case is the instance-based individual representation discussed in section 8.3. Recall that in this case each gene usually corresponds to a data instance, and the gene's value explicitly represents the index of a cluster to which the corresponding instance is assigned. For instance, if the third gene has the value 2, this means that the third data instance is assigned to cluster 2. Mathematically: $M_{32} = 1$ and, for all $j \neq 2$, $M_{3j} = 0$.

If a centroid/medoid-based individual encoding is used, as discussed in section 8.2, each data instance is assigned to its nearest cluster. Of course, this requires that the algorithm computes the distance between each data instance and each of the cluster centroids according to a given measure, typically the Euclidean distance or the Manhattan distance.

Note that the constraint that each data instance is assigned to exactly one cluster is straightforwardly satisfied by the two above-mentioned individual encoding/decoding methods. However, when using a cluster description-based individual encoding (section 8.1), the satisfaction of that constraint may require some special-purpose operator (assuming that we do want to satisfy that constraint, which will not always be the case).

Once an individual is decoded and the partitioning matrix M is computed, the fitness of an individual (a candidate clustering solution) is usually computed based on some of the principles of clustering evaluation discussed in section 2.4 (see also subsection 9.1.5). In particular, the fitness function should favor solutions with small intra-cluster (within-cluster) distance and large inter-cluster (between-cluster) distance [Krishma and Murty 1999; Kim et al. 2000; Hall et al. 1999].

For instance, one possible fitness function is [Hall et al. 1999]:

$$\text{Fit} = \sum_{i=1}^{n} \min\{D_{i1}, D_{i2}, ... , D_{iK}\}$$

where n is the number of data instances to be clustered, K is the number of clusters, and D_{i1} ... D_{iK} are the distances between the i-th data instance and each of the K cluster centroids. This formula computes the summation of the distance between every data instance and the centroid of the cluster to which that instance is allocated. This is equivalent to computing the summation of intra-cluster (within-cluster) distance for all clusters. (The use of this formula implicitly assumes that the number of clusters is fixed for all individuals.) Hall et al. have extended the above formula to penalize degenerate solutions – see below.

8.4.1 Coping with Degenerate Partitions

Let K be the number of clusters associated with an individual of an EA. One can say that an individual represents a degenerate partition if one or more of its clusters are empty, i.e., no data instance belongs to those clusters. In this case the number of effective clusters is smaller than K, since it does not make sense to return an empty cluster for the user. Some ways of coping with the degenerate-partition problem are briefly reviewed next.

[Hall et al. 1999] observed that their GA (discussed in section 8.2.1) occasionally got caught in a local minimum associated with a degenerate partition. To minimize the chance of the GA getting trapped at a degenerate partition, the authors have added to the fitness function a penalty term whose value is proportional to the number of empty clusters. More precisely, let Fit be the original value of an individual's fitness. The new fitness value is given by: $Fit = (1 + e) \times Fit$, where e is the number of empty clusters, which can vary in the range $0...K$. Since the GA's goal is to minimize the value of the fitness function, the above formula effectively penalizes candidate solutions having degenerated partitions, as desired.

[Meng et al. 2000] used a small variation of the previous formula to penalize degenerate partitions, namely: $Fit = Fit \times (1 + e / K)$ where e is the number of empty clusters and K is the number of clusters.

Adding a penalty term to the fitness function is not the only way to cope with the degenerate-partition problem. In [Krishma and Murty 1999] an individual representing a partition with empty clusters is deemed an illegal individual. They convert illegal individuals to legal ones by creating the necessary number of new singleton clusters. This is done by placing in each empty cluster a data instance i from the cluster c with the maximum within-cluster variation. i is the farthest instance from the center of the cluster c.

In addition, in [Ghozeil and Fogel 1996] hyperboxes (cluster descriptions) that contain no data can be pruned at the completion of a clustering run. Such a method could also be used during the EA run.

8.5 EAs vs Conventional Clustering Techniques

Most heuristic techniques for clustering perform a local search in the space of candidate solutions. These techniques often rely on local heuristics applied to individual data points (e.g., nearest-neighbor methods). This situation is similar to the use of local heuristics in prediction-rule discovery.

Hence, at a high level of abstraction, the motivation for using EAs in clustering is conceptually similar to the motivation for using EAs in prediction-rule discovery. EAs usually perform a global search in the candidate solution space, and they tend to cope better with attribute interaction.

The search space associated with a clustering problem tends to be very large and rugged – with many local optima. Therefore, one would normally expect the use of EAs to be more effective than the use of techniques based on local heuristics. Of course, this expected advantage is obtained at the expense of an expected increase in processing time. In general EAs are slower than conventional, local-heuristic clustering techniques.

There is some evidence that EAs for clustering do outperform the classical K-means algorithm [Meng et al. 2000; Krishma and Murty 1999; Babu and Murty 1993]. In particular, [Krishma and Murty 1999] and [Babu and Murty 1993] report that the difference of performance between an EA and K-means tends to increase as the number of clusters is increased.

However, most projects evaluate the performance of an EA for clustering in just a few (typically less than five) data sets, so that these results are not entirely conclusive. Furthermore, in some projects – such as [Srikanth et al. 1995] and [Ghozeil and Fogel 1996] – the algorithm was applied only to two-dimensional data. In addition, in the literature EAs are often compared with multiple runs of K-means, each with a different randomly-chosen initial partition. It would be more challenging to compare the evolutionary algorithm with another heuristic search method (say simulated annealing or tabu search) designed to optimize the initial partition of K-means.

References

[Babu and Murty 1993] G.P. Babu and M.N. Murty. A near-optimal initial seed selection in k-means algorithm using a genetic algorithm. *Pattern Recognition Letters, 14,* 763–769, 1993.

[Backer 1995] E. Backer. *Computer-Assisted Reasoning in Cluster Analysis.* Prentice-Hall, 1995.

[Deb 2000] K. Deb. Encoding and decoding functions. In: T. Back, D.B. Fogel and Z. Michalewicz (Eds.) *Evolutionary Computation 2: Advanced Algorithms and Operators,* 4–11. Bristol, Institute of Physics Publishing, 2000.

[Estivill-Castro and Murray 1997] V. Estivill-Castro and A.T. Murray. Spatial clustering for data mining with genetic algorithms. *Tech. Report FIT-TR-97-10*. Queensland University of Technology. Australia. 1997.

[Falkenauer 1998] E. Falkenauer. *Genetic algorithms and grouping problems.* Wiley, 1998.

[Ghozeil and Fogel 1996] A. Ghozeil and D.B. Fogel. Discovering patterns in spatial data using evolutionary programming. *Genetic Programming 1996: Proceedings of the 1st Annual Conference,* 521–527. MIT Press, 1996.

[Hall et al. 1999] L.O. Hall, I.B. Ozyurt, J.C. Bezdek. Clustering with a genetically optimized approach. *IEEE Transactions on Evolutionary Computation 3(2)*, 103–112. 1999.

[Kim et al. 2000] Y. Kim, W.N. Street and F. Menczer. Feature selection in unsupervised learning via evolutionary search. *Proceedings of the 6th ACM SIGKDD International Conference on Knowledge Discovery and Data Mining (KDD '2000)*, 365–369. ACM, 2000.

[Krsihma and Murty 1999] K. Krsihma and M. N. Murty. Genetic k-means algorithm. *IEEE Transactions on Systems, Man and Cybernetics - Part B: Cybernetics,* 29(3), 433–439. June 1999.

[Krzanowski and Marriot 1995] W.J. Krzanowski and F.H.C. Marriot. *Kendall's Library of Statistics 2: Multivariate Analysis – Part 2. Chapter 10 – Cluster Analysis,* 61–94. London: Arnold, 1995.

[Lozano and Larranaga 1999] J.A. Lozano and P. Larranaga. Applying genetic algorithms to search for the best hierarchical clustering of a dataset. *Pattern Recognition Letters 20,* 911–918, 1999.

[Meng et al. 2000] L. Meng, Q.H. Wu and Z.Z. Yong. A faster clustering algorithm. *Real-World Applications of Evolutionary Computing (Proceedings of the Evoworkshops 2000). Lecture Notes in Computer Science 1803,* 22–33. Springer, 2000.

[Park and Song 1998] Y. Park and M. Song. A genetic algorithm for clustering problems. *Genetic Programming 1998: Proceedings of the 3rd Annual Conference,* 568–575. Morgan Kaufmann, 1998.

[Rizzi 1998] S. Rizzi. Genetic operators for hierarchical graph clustering. *Pattern Recognition Letters 19,* 1293–1300, 1998.

[Srikanth et al. 1995] R. Srikanth, R. George, N. Warsi, D. Prabhu, F.E. Petry, B.P. Buckles. A variable-length genetic algorithm for clustering and classification. *Pattern Recognition Letters 16(8)*, 789–800. 1995.

9 Evolutionary Algorithms for Data Preparation

> "In reality, the boundary between pre-processor and classifier is arbitrary. If the pre-processor generated the predicted class label as a feature [attribute], then the classifier would be trivial. Similarly, the pre-processor could be trivial and the classifier do all the work."
>
> [Sherrah et al. 1997, p. 305]

Clearly the quality of discovered knowledge strongly depends on the quality of the data being mined. This has motivated the development of several algorithms for data preparation tasks, as discussed in chapter 4.

This chapter discusses several Evolutionary Algorithms (EAs) for data preparation tasks. More precisely, this chapter is divided into five sections, as follows. Section 9.1 discusses EAs (mainly Genetic Algorithms (GAs)) for attribute selection. Section 9.2 discusses EAs for attribute weighting. Section 9.3 discusses a GA for combining attribute selection and attribute weighting. Section 9.4 discusses Genetic Programming (GP) for attribute construction. Section 9.5 discuss a hybrid GA/GP algorithm for combining attribute selection and attribute construction.

It should be noted that, although this chapter discusses EAs for data preparation as a pre-processing step for data mining, EAs can also be used for post-processing the knowledge discovered by a data mining algorithm – see, e.g., [Thompson 1998, 1999], where a GA optimizes the voting weights of classifiers in an ensemble of classifiers.

Unless mentioned otherwise, throughout this chapter we will assume that the target data mining task (which will use the results of the data preparation task) is classification. This is the data mining task most addressed in the literature, and most data-preparation evolutionary algorithms were developed for this task. One exception is subsection 9.1.5, which addresses attribute selection for the data mining task of clustering.

9.1 EAs for Attribute Selection

As discussed in section 4.1, attribute selection is a data preparation task that consists of selecting, among all available attributes, a subset of attributes relevant for the target data mining task.

This section is divided into six parts. In subsection 9.1.1 we discuss how to encode candidate solutions – selected attribute subsets – into a GA individual, as well as some genetic operators appropriate for the corresponding individual encoding. In subsection 9.1.2 we discuss the main issues involved in the design of fitness functions for GAs for attribute selection. In subsection 9.1.3 we discuss attribute selection via search for partially-defined instances. In subsection 9.1.4 we discuss simultaneous attribute selection and instance selection. The above-mentioned subsections focus on attribute selection for the classification task. In subsection 9.1.5 we discuss attribute selection for the clustering task. Finally, subsection 9.1.6 concludes this section with some comments on the effectiveness of EAs for attribute selection.

9.1.1   Individual Encoding
and Genetic Operators

9.1.1.1 The Standard Individual Encoding
for Attribute Selection

The search space of an attribute selection problem consists of all possible attribute subsets. Each state in this search space can be represented by a fixed-length binary string containing m bits, where m is the number of available attributes. The i-th bit, $i=1,...,m$, indicates whether or not attribute A_i is selected. For instance, let $m = 10$. Then the binary string 0 1 0 1 0 0 0 0 0 1 corresponds to an attribute subset – a candidate solution – where only attributes A_2, A_4 and A_{10} are selected.

This candidate-solution encoding naturally suggests the use of a simple genetic algorithm (GA) for attribute selection. By simple GA we mean a GA with fixed-length, binary string representation, in the style of the GA popularized by [Goldberg 1989]. We will refer to the use of a simple GA with binary, fixed-length string for attribute selection, as described above, as the "standard" individual encoding for attribute selection.

The main advantage of this standard encoding is its simplicity. Actually, some authors have emphasized that when using this approach there is no need to develop problem-dependent genetic operators. Any "standard" crossover and mutation operators developed for fixed-length binary strings will do.

It should be noted, however, that the fact that standard genetic operators can be naturally used does not imply that they are the best choice of operators to be used with the standard individual encoding for attribute selection. Actually, some authors have challenged the effectiveness of standard genetic operators for this kind of encoding [Chen et al. 1999; Guerra-Salcedo et al. 1999], as discussed in the following.

The authors start arguing that in attribute selection the "1"s of an individual may be more informative than its "0"s. The occurrence of an attribute (indicated by "1" in the corresponding gene) in two individuals being mated is a good indicator that the attribute is relevant. However, the non-occurrence of an attribute (indicated by "0" in the corresponding gene) may be inappropriate to hypothesize that the attribute is entirely irrelevant, since the attribute can be "weakly relevant".

To find weakly relevant attributes the authors use the heuristic of estimating which attributes are strongly relevant and start the search from this subset. Therefore, they propose a commonality-based crossover for the attribute selection task. Selected attributes (with "1" in the corresponding genes) are chosen as the basis of commonality between individuals. The partial solutions consisting of the common "1"s are then used to initialize a probabilistic search method (based on multiple bit mutations), which is used to search for additional (weakly relevant) attributes.

Another form of commonality-based crossover for attribute selection is discussed in [Emmanouilidis et al. 2000]. In this work a commonality-based crossover is used with a multi-objective evolutionary algorithm (MOEA). (The basic idea of MOEAs was briefly discussed at the end of section 6.5. In addition, a MOEA for attribute selection will be discussed in subsection 9.1.5.)

9.1.1.2 An Attribute Index-Based Individual Encoding for Attribute Selection

Although the above standard individual encoding for attribute selection is by far the most used, other forms of encoding a selected attribute subset into a GA individual are also possible, and may have some advantages.

An alternative form of individual encoding for attribute selection was proposed by [Cherkauer and Shavlik 1996]. In this case an individual consists of N genes, where each gene can contain either the index (id) of an attribute or a flag – say 0 – indicating no attribute. The value of N is a user-specified parameter. An attribute is considered selected if its corresponding index occurs in at least one of the genes of the individual. For instance, consider the following individual, where $N = 5$:

$$0 \ \ A_1 \ \ A_4 \ \ 0 \ \ A_1.$$

In this case only attributes A_1 and A_4 are selected. The fact that A_1 occurs twice in the individual, whereas A_4 occurs only once, is irrelevant for the purpose of decoding the individual into a subset of selected attributes.

Two motivations for this unconventional individual encoding are as follows. First, the fact that an attribute can occur more than once in an individual can act as a redundancy mechanism that increases robustness and slows the loss of genetic diversity. Second, in this encoding the length of an individual is independent of the number of original attributes. Hence, this approach is more scalable to data sets with a very large number of attributes. In this case we can specify a number of genes much smaller than the number of original attributes, which would not be possible with the standard binary encoding for attribute selection.

The use of such an unconventional individual encoding suggests the development of new genetic operators tailored for this kind of encoding. Hence, in addition to crossover and mutation operators, the proposed GA uses a genetic operator called *Delete Attribute* (or *Delete Feature*, in the author's terminology). This operator accepts as input one parent and produces as output one offspring where all occurrences of one (equiprobably chosen) attribute are removed from the parent. Therefore, this operator has a bias that favors the selection of smaller attribute subsets. For instance, in the above example of individual, if attribute A_1 is chosen to be deleted then the individual will become: 0   0 A_4 0   0.

9.1.2   Fitness Evaluation

As we discussed in chapter 4, attribute selection methods can be broadly classified into wrapper and filter approaches. A key difference between these two approaches is that a measure of performance of a classifier, run using only the selected attributes, is used in the wrapper approach, but not in the filter approach. In other words, in the wrapper approach the fitness function of a GA for attribute selection involves a measure of performance of a classification algorithm using only the subset of attributes selected by the corresponding GA individual. This basic idea is illustrated in Figure 9.1.

According to this simple definition, the vast majority of GAs for attribute selection follow the wrapper approach.

There is, however, an alternative view, in which a GA often has aspects of both wrapper and filter approaches, depending on what is measured in the fitness function. If the fitness function involves only a measure of performance of the classifier – e.g., the classification accuracy rate – then the GA is definitely following a wrapper approach.

Now, suppose the fitness function involves both the classifier's accuracy rate and the number of selected attributes. The value of this latter criterion is independent of the classifier. It depends only on the genes of an individual. Hence, the use of this criterion adds to the fitness function a certain flavor of the filter approach, even though one might argue that the wrapper criterion of classifier's accuracy rate is still predominant in most cases. The predominance of the wrapper criterion in most cases is normally due to the use of a fitness function which is a weighted sum of classification accuracy rate and number of selected attributes, where the weight of the former term is usually greater than the weight of the latter.

In any case, the distinction between wrapper and filter approaches gets more fuzzy if the fitness function includes another filter-oriented criterion besides the number of selected attributes. An example is the fitness function for attribute selection proposed by [Bala et al. 1996]. This function takes the form:

$$Fitness(s) = Info(s) - Cardinality(s) + Accuracy(s),$$

where s is the candidate attribute subset associated with an individual, *Accuracy(s)* is the classification accuracy rate of a classification algorithm using only the attributes in s, *Cardinality(s)* is the number of attributes in s, and *Info(s)* is an information-theoretic measure which is an estimate of the discriminatory power of the attributes in s. Note that *Cardinality(s)* and *Info(s)* take on values independent of the classification algorithm used, so that they are filter oriented criteria.

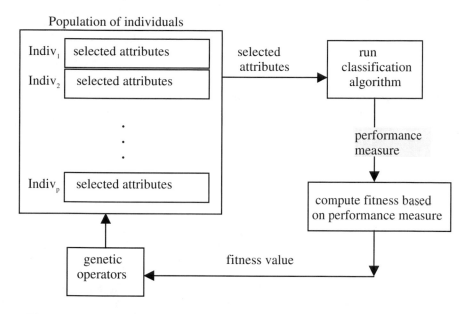

Figure 9.1: Fitness evaluation in the wrapper approach for attribute selection

It is also possible to define a fitness function that follows a pure filter approach, ignoring altogether the classification accuracy rate of the classifier. This approach normally has the advantage of reducing the processing time of the GA. This stems from the fact that in general the computation of a filter-oriented criterion for an attribute subset is considerably faster than running a classification algorithm with that attribute subset, as required in the wrapper approach.

One example of a GA following the filter approach can be found in the above-mentioned work of [Bala et al. 1996]. In one of the experiments reported by the authors the wrapper component of the fitness function, Accuracy(s), was switched off, so that the GA effectively followed a purely-filter approach. Unfortunately, this purely-filter variant did not produce good results.

[Chen and Liu 1999] propose an adaptive algorithm for attribute selection following the filter approach. A candidate solution is represented by the standard binary encoding for attribute selection (subsection 9.1.1.1). The algorithm iteratively generates a candidate attribute subset and evaluates the quality of that subset according to a filter-oriented function – see below. In the first iteration the generation of candidate attribute subsets is random and unbiased. However, as more and more candidate attribute subsets are generated and evaluated, the algorithm uses the result of those evaluations to modify the probability that an attribute is included in the next candidate attribute subset. Hence, if an attribute occurs in many good attribute subsets its probability of being included in the next candidate attribute subset to be generated is increased.

The proposed algorithm is better described as an adaptive algorithm, rather than an evolutionary one, because the generation of new solutions during the search is not directly based on existing solutions. There is no selection, crossover or mutation operator.

In any case we decided to mention this algorithm in this section because of its "fitness function" (loosely speaking), which seems a promising fitness function to be used in a filter approach-based GA for attribute selection. This function follows a filter criterion based on the notion of inconsistency rate [Wang and Sundaresh 1998] (see also subsection 4.1.2). The basic idea is that two data instances are considered inconsistent if their values are the same (they match) for all the attributes in the selected attribute subset but they have different classes. The general goal is to find the smallest subset of attributes that can maintain the lowest inconsistency rate.

9.1.2.1 *Interactive Fitness Evaluation*

As mentioned above, when using a GA following the wrapper approach the quality of a candidate attribute subset is evaluated by some measure of performance of a classification algorithm, run using only that attribute subset. So far we have assumed that this measure of performance is some objective, automatically-computed measure, such as classification accuracy rate.

Although this objective approach for measuring performance is by far the most used, it turns out that one can instead use a subjective measure of performance of the classification algorithm, as interactively evaluated by the user along evolution.

An interactive GA for attribute selection is described in [Terano and Ishino 1998]. In essence, each individual is a binary string representing a candidate attribute subset, as in the standard individual encoding described in section 9.1.1.1. In order to evaluate individuals, their attribute subsets are given to a classification algorithm, which uses those subsets to generate sets of classification rules. Then a user interactively and subjectively selects good rules and rule sets. The user is supposed to be a domain expert, so that (s)he can evaluate the rules according to a number of subjective criteria that could hardly be incorporated into a fully-automated system. Then the individuals having attributes occurring in the selected rules or rule sets are selected as parents for producing new offspring. In the experiments reported by the authors, the domain expert has the following profile and role [Terano and Ishino 1998, p. 399]:

"She knows the basic principles of inductive learning programs, statistical techniques, and is able to understand the output results. Using the output forms of decision trees and corresponding rule sets from C4.5 programs, she has been required to interactively and subjective evaluate the quality of the acquired knowledge from the viewpoints of simplicity, understandability, accuracy, reliability, plausibility, and application of the knowledge."

The motivation for this approach is that, since it puts the user in the loop, there is a better chance of discovering rules that are useful for the user. For instance, the authors report that in one experiment, involving marketing decisions about oral care products, the system successfully discovered easy-to-understand, good rules – as judged by a human expert.

The advantage of this customization is obviously achieved at the expense of a reduction in the autonomy and speed of the system. In practice, to limit the time spent by the user on classifier-performance evaluation, one often has to use a GA with a small number of individuals and a small number of generations.

9.1.2.2 Summary of Fitness Functions Involving a Wrapper Criterion

The vast majority of GAs for attribute selection follow the wrapper approach, or a hybrid wrapper/filter approach, as discussed above. A summary of the main aspects of fitness functions for attribute selection involving a wrapper criterion is presented in Table 9.1. In general, each row of this table corresponds to one algorithm. The first column of this table contains the bibliographical reference describing the corresponding algorithm. The second column indicates the kind of classification algorithm used, following a wrapper approach. The third column describes, at a high level of abstraction, the main criteria considered by the fitness function of the GA. An additional overview of GAs for attribute selection can be found in [Martin-Bautista and Vila 1999] and [Freitas 2002].

In the above discussion we have mentioned only the classification accuracy rate as an example of a wrapper-oriented criterion for attribute selection. Although this is by far the most used criterion, other wrapper-oriented criteria could also be used. One possibility is suggested by the fact that some classifiers' output provide potentially useful information about the relative importance of each attribute. In principle this kind of information could be fedback to the GA. For instance, as pointed out by [Bala et al. 1995], the relative position of an attribute in a decision tree is an indicator of its relevance, i.e., attributes at shallower levels (closer to the root) of the tree tend to be more relevant than attributes at deeper tree levels.

Table 9.1: Main aspects of fitness functions of GAs for attribute selection

Reference	Classification algorithm	Criteria used in fitness function
[Bala et al. 1995]	decision tree	predictive accuracy, No. of selected attributes
[Bala et al. 1996]	decision tree	predictive accuracy, info. content, No. of selected attributes
[Chen et al. 1999]	euclidean decision table	based first on predictive accuracy, and then on No. of selected attributes

Table 9.1: (Continued)

Reference	Classification algorithm	Criteria used in fitness function
[Guerra-Salcedo and Whitley 1998]	euclidean decision table	predictive accuracy
[Guerra-Salcedo et al. 1999]	euclidean decision table	predictive accuracy
[Cherkauer and Shavlik 1996]	decision tree	predictive accuracy , No. of selected attributes, average decision-tree size
[Terano and Ishino 1998]	decision tree	subjective evaluation, predictive accuracy, rule set size
[Vafaie and DeJong 1998]	decision tree	predictive accuracy
[Yang and Honavar 1997, 1998]	neural network	predictive accuracy, attribute cost
[Moser and Murty 2000]	nearest neighbor	predictive accuracy, No. of selected attributes
[Ishibuchi and Nakashima 2000]	nearest neighbor	predictive accuracy, No. of selected attributes, No. of selected instances (attribute and instance selection)
[Emmanouilidis et al. 2000]	neural network	predictive accuracy, No. of selected attributes (multi-objective evaluation)

Finally, it should be noted that so far our discussion has focused on GAs for selecting attributes for a single classification algorithm, or classifier. However, a GA can also be used for selecting attributes that will be used for generating an ensemble of classifiers. In the last few years the use of an ensemble of classifiers (e.g., an ensemble of rule sets), rather than a single classifier (e.g., a single rule set), has become popular in the machine learning literature. This is the case particularly when the primary goal of the system is to maximize predictive accuracy. (The advantage of using an ensemble of classifiers seems less clear when discovered-knowledge comprehensibility is important, since a relatively large number of classifiers can hinder the interpretation and validation of the classifiers by the user.)

A GA that performs attribute selection for generating an ensemble of classifiers is discussed in [Guerra-Salcedo and Whitley 1999a, 1999b]. The basic idea is that a conventional GA for attribute selection is run many times, and each run selects a subset of attributes. Then each of the selected attribute subsets is given for a classifier of the ensemble, so that different classifiers of the ensemble are trained with different attribute subsets.

9.1.3 Attribute Selection via Search
for Partially Defined Instances

A different approach for performing attribute selection is suggested by [Llora and Garrel 2001]. In this approach each individual of a GA represents a set of partially-defined instances. An instance is partially-defined if at least one of its predictor attributes has a known value. Hence, an original, fully-defined instance (having known values for all predictor attributes) can be transformed into a partially-defined instance by discarding some of its attributes. For instance, suppose that a data set has 3 predictor attributes, denoted A_1, A_2 and A_3, and a goal attribute G. An example of a fully-defined instance is $<A_1 = 3, A_2 = 2, A_3 = 4, G = yes>$, while examples of partially-defined instances are $<A_2 = 2, G = yes>$ and $< A_1 = 3, A_2 = 2, G = yes>$.

Hence, unlike a conventional attribute selection approach, where all instances end up containing the same set of selected attributes, in this approach different instances end up containing different selected attributes. This approach was used to select partially-defined instances for a nearest neighbor algorithm.

9.1.4 Joint Attribute Selection
and Instance Selection

In the previous subsections we have seen how a simple GA can be used for attribute selection. The problem of attribute selection can be regarded as a particular kind of data selection (or data reduction) problem. More generally, one can reduce the amount of data being mined by selecting not only a set of attributes but also a set of data instances, out of all original attributes and data instances. This is a complex problem, and we will discuss here just one evolutionary algorithm designed to solve it.

[Ishibuchi and Nakashima 2000] propose a GA for simultaneous attribute selection and instance selection. The GA is used as a wrapper around a classification algorithm, more precisely a nearest neighbor algorithm (section 3.3). The basic idea is similar to the use of a GA as a wrapper in attribute selection, which was discussed in detail in the previous subsections. The main difference is that now a candidate solution encoded into a GA individual consists of both selected attributes and selected instances, rather than just selected attributes. The data (attributes and instances) selected by an individual is given as input to a classification algorithm, and the performance of that algorithm is fedback to the GA via computation of a fitness function (discussed below), as usual in the wrapper approach.

To encode a candidate solution into an individual the authors use a simple binary string of length $m + n$, where m is the number of original attributes and n is the number of original data instances. Hence, an individual is composed of two binary substrings, as illustrated in Figure 9.2. The i-th gene, $i=1,...,m$, can take on the value 1 or 0 indicating whether or not the i-th attribute is selected. Similarly, the $(m + j)$-th gene, $j=1,...,n$, can take on the value 1 or 0 indicating whether or

genes specifying selected attributes   genes specifying selected instances

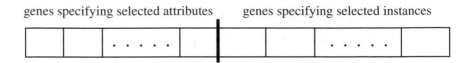

Figure 9.2: Individual representing both selected attributes and selected instances

not the j-th data instance is selected. Since this is a straightforward extension of the standard individual encoding for attribute selection (section 9.1.1.1), we can call it the standard individual encoding for simultaneous attribute and instance selection.

This form of individual encoding has the advantage of being simple, but it has the disadvantage of not being scalable for large data sets, since an individual's length is directly proportional to the number of data instances, n. In the case of large data sets, where the value of n is large, the use of this encoding would tend to render the GA too slow, and one alternative individual encoding would probably be necessary – at least for the substring encoding the selected data instances.

Ishibuchi and Nakashima also proposed a biased mutation for the binary substring encoding the selected data instances. More precisely, for the n bits corresponding to the n data instances, the probability of mutating a 1 into a 0 is larger than the probability of mutating a 0 into a 1. This strategy favors the generation of individuals with fewer selected data instances. The authors use this biased mutation only for the instance-selection substring, and not for the attribute-selection substring. However, this biased mutation could be used for both substrings, if desired. In this case the use of a biased mutation in the attribute-selection substring would favor the selection of fewer attributes. This could be helpful particularly in data sets with a large number of attributes, where many (perhaps most) attributes are expected to be irrelevant.

In any case, the GA proposed by the authors also uses a fitness function that favors individuals that not only have a high predictive accuracy, but also select few attributes and few instances. These three criteria are combined into a weighted sum as follows:

$$Fitness = w_{acc} \cdot Acc - w_{att} \cdot |A| - w_{inst} \cdot |I| ,$$

where Acc is a measure of predictive accuracy (the number of correctly classified instances) of the classification algorithm, run with only the attributes and instances selected by the individual; $|A|$ is the number of attributes selected by the individual; $|I|$ is the number of instances selected by the individual; and w_{acc}, w_{att} and w_{inst} are user-defined weights.

9.1.5 Attribute Selection for Clustering

In the previous subsections we discussed several kinds of GA for attribute selection. However, all of those algorithms had one characteristic in common: they were designed for the classification task. In this subsection we address the use of Evolutionary Algorithms (EAs) for attribute selection in the clustering task.

First of all, note that in both tasks – classification and clustering – a candidate solution has essentially the same nature, i.e., a subset of attributes, out of all available attributes. Therefore, the basic ideas of individual encoding and genetic operators discussed in the previous subsections (in the context of classification) can be used in the case of clustering as well. The main difference is that, of course, the fitness function must be very different in the two cases.

We now briefly review an EA that performs attribute selection for clustering [Kim et al. 2000]. In this work an EA is used as a wrapper around a K-means algorithm. (The K-means algorithm was briefly reviewed in subsection 2.4.2.) This algorithm is an agent-based EA that uses a local selection schema [Menczer et al. 2000]. In addition, it uses only mutation to explore the search space – no crossover is used. Here we will focus on the individual encoding and fitness function of the algorithm, rather than on its selection schema or agent-based aspects.

Each individual, or agent, of the EA represents both an attribute subset and a value of K (the number of clusters). The individual encoding consists of two parts, each of them a fixed-length binary substring. The first substring implements the standard individual encoding for attribute selection – each of those bits corresponds to an attribute, which is considered selected or not depending on whether the bit is "1" or "0", respectively, as described in subsection 9.1.1.1. The second substring is used to encode the number of clusters, K.

When an individual is decoded its attribute subset and its value of K are given to a K-means algorithm. A measure of K-means' performance is then fedback to the EA, as part of the fitness function. More precisely, the fitness function consists of four measures, as follows:

(a) a measure of the cohesiveness of clusters, favoring dense clusters;
(b) a measure of the distance between clusters and the global centroid, which favors well-separated clusters;
(c) a measure of the "simplicity" of the number of clusters, which favors candidate solutions with a smaller number of clusters;
(d) a measure of the "simplicity" of the selected attribute subset, which favors candidate solutions with a small number of selected attributes.

Note that measures (a) and (b) are derived from the result produced by the K-means algorithm, characterizing the wrapper approach, whereas measures (c) and (d) are independent of the result of K-means, characterizing the filter approach. Therefore, this fitness function actually follows a hybrid wrapper/filter approach.

Each of the above measures is normalized into the unit interval and is considered as an objective (to be maximized) in a multi-objective optimization problem. In other words, the EA is a multi-objective algorithm, in the sense that it searches for non-dominated solutions (the Paretto front) taking into account all the objectives. Recall that a solution S_1 is said to dominate a solution S_2 if the following two conditions hold:

(a) S_1 is better (it has a higher measure) than or equal to S_2 in all the objectives;
(b) S_1 is better than S_2 in at least one of the objectives.

A solution S is said to be non-dominated if no solution dominates S.

9.1.6 Discussion

From the discussion of the previous subsections one can see that several EAs for attribute selection have been proposed. Most of these EAs are relatively simple GAs, often using a binary encoding, and most of them follow the wrapper approach, often using hybrid wrapper/filter criteria in the fitness function.

It is interesting to compare EAs with non-evolutionary methods for attribute selection. As pointed out by [Yang and Honavar 1997, p. 381], many of the non-evolutionary attribute selection methods proposed in the literature have one (or both) of the following disadvantages. First, they often assume monotonicity of the measure used to evaluate the quality of a candidate attribute subset, typically a measure of predictive accuracy in the wrapper approach. The monotonicity assumption – adding attributes to an attribute subset is guaranteed not to reduce predictive accuracy – is not very realistic in the wrapper approach, due to attribute interactions, noisy attributes, etc. Second, many non-evolutionary attribute selection methods were not designed to cope with multiple criteria for evaluating a candidate attribute subset, such as predictive accuracy, number of selected attributes, etc.

Intuitively, EAs can be used to try to avoid these problems. EAs seem to be suitable for attribute selection, due to their ability in performing a global search and coping better with attribute interaction than "conventional" sequential, local-search attribute-selection methods. They make no assumption about monotonicity of the fitness function, which is the function used to evaluate the quality of a candidate attribute subset. In addition, GAs are flexible enough to naturally allow the evaluation of a candidate attribute subset to be based on a combination of different criteria.

Concerning empirical evaluations of EAs for attribute selection, in some projects the performance of the EA is compared only with the performance of a classification algorithm using all original attributes. A more challenging test for an EA performing attribute selection is to compare the EA with another attribute selection method.

For instance, [Sharpe and Glover 1999] compared a GA following the wrapper approach with forward sequential selection (subsection 4.1.2) – which is a greedy search method – across five data sets. Overall (but not always) the GA achieved better results with respect to predictive accuracy.

A more extensive comparison of a GA for attribute selection with non-evolutionary attribute selection methods, following the wrapper approach, is reported in [Kudo and Skalansky 2000]. More precisely, the authors compared a

GA with 14 non-evolutionary attribute selection methods (some of them variants of each other) across 8 different data sets.

Two limitations of this study, from a data mining viewpoint, are that the only classification algorithm used in the experiments was a nearest neighbor (1-NN) classifier and that the only criterion used for evaluating a candidate attribute subset was predictive accuracy. Hence, there was no emphasis on the discovery of high-level, comprehensible prediction rules. In any case, this study produced several interesting results concerning the relative effectiveness of several different attribute selection methods with respect to the maximization of predictive accuracy.

One of the main conclusions of this study was that sequential floating search methods, which are very respectable attribute selection methods, are suitable for small- and medium-scale attribute-selection problems, whereas GAs are suitable for large-scale problems (where the number of attributes is over 50) [Kudo and Sklansky 2000, p. 25].

In data sets with a large number of attributes, the advantage of the global, parallel search of the GA over the local, sequential search performed by other attribute selection methods is clear in the following quote from [Kudo and Sklansky 2000, p. 39]:

"Even if a particular GA takes much more time than SFFS and SBFS [two sequential floating search methods], the GA has a high possibility to find better solutions than the two algorithms. This is because SFFS and SBFS do not have a mechanism of jumping from a subset to another very different subset. They trace a sequence of subsets of which adjacent subsets are different by only one feature [attribute]."

However, one should bear in mind that GAs following the wrapper approach usually have a considerably long processing time. This raises the question of whether it is better to run a GA or, alternatively, to use the time that would be taken by the GA to run several faster attribute selection methods and pick up the best attribute subset selected by all those methods [Freitas 1997] – see also subsection 4.1.2.1. This question (which has been little explored so far) involves a measure of the cost-effectiveness of a GA for attribute selection, rather than just a measure of its effectiveness.

9.2 EAs for Attribute Weighting

As discussed in subsection 4.1.3, attribute selection can be regarded as a particular case of attribute weighting. In attribute selection we can assign just two "weights" to each attribute – either 0, indicating that the attribute is not selected, or 1, indicating that the attribute is selected – while in attribute weighting the weights can take on any real value in a desired range, say the [0..1] continuum.

9.2.1 Attribute Weighting

[Punch et al. 1993] proposed a GA for attribute weighting. The proposed algorithm works as a wrapper around an attribute-weighted k-nn (nearest neighbor)
algorithm. Similarly to the use of a GA as a wrapper in attribute selection, the use
of a GA as a wrapper in attribute weighting involves two basic ideas: (a) an individual specifies a value for each of the attribute weights; (b) the fitness function
involves a measure of the predictive accuracy of a k-nn algorithm using the attribute weights specified by the individual. This basic idea is illustrated in Figure
9.3.

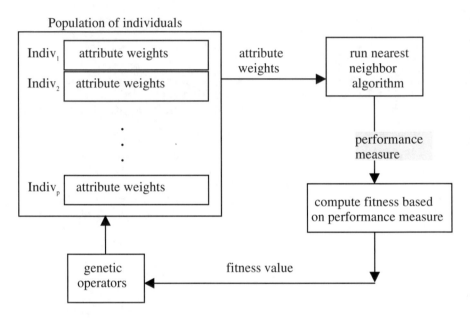

Figure 9.3: Overview of the use of a GA as a wrapper in attribute weighting

In order to encode attribute weights into a GA individual, Punch et al. use a
binary, fixed-length encoding. Each attribute weight is allocated a fixed number
of bits in the genome, and the total length of an individual is simply the product
of that number of bits by the number of attributes.

In their experiments the authors observed that, as evolution goes on, some attribute weights approach 0 – indicating that the attribute is irrelevant – but very
rarely are effectively set to 0. To simplify the run of the k-nn algorithm, if an
attribute weight falls below a small pre-specified threshold value then the attribute weight is effectively set to 0.

After experimenting with several different fitness functions they advocate this
one:

$$\text{Fitness} = w_1 \, (TotPred - CorrectPred) \,/\, TotPred \,+\, w_2 \,(nmin \,/\, k) \,/\, TotPred,$$

where *CorrectPred* and *TotPred* are the number of correct predictions and the total number of predictions, respectively, k is the number of nearest neighbors used for classifying a data instance, and *nmin* is the number of neighbors in the set of k nearest neighbors that have a minority class, i.e., a class different from the predicted (majority) class. For instance, suppose $k = 5$, three of the k nearest neighbors have class c_1 and the other two have class c_2. In this case the predicted class is c_1, and $nmin = 2$. w_1 and w_2 are user-defined weights. This fitness function favors candidate solutions that not only minimize the number of incorrect predictions, but also minimize the value of *nmin*. The authors suggest the values of 2 and 0.2 for the weights w_1 and w_2, respectively.

9.2.2 Towards Constructive Induction

The above-mentioned work of [Punch et al. 1993] also extended an individual genome for attribute weighting with some special fields representing a simple form of attribute construction. The basic idea is that the genome is extended with some triples of the form: $<Index_1, Index_2, W>$, where $Index_1$ and $Index_2$ are attribute indexes identifying a pair of attributes and W is the weight assigned to the new attribute that is constructed by applying a given operation to the attributes identified by $Index_1$ and $Index_2$. In the experiments reported by the authors this operation is simply the product of the values of the attributes identified by $Index_1$ and $Index_2$. This implicitly assumes that the attributes are numerical. In the case of categorical attributes, such as *Sex* or *Marital_Status,* another kind of meaningful operator would have to be defined – e.g., logical *AND* and *OR* operators.

By including a relatively small number of these triples in the genome the GA performs a heuristic search for good pairs of attributes, avoiding the need for exhaustively evaluating all possible attribute pairs. Note that by extending an individual genome in this fashion it is possible that a given original attribute be assigned a low weight, for being weakly correlated with the goal attribute, yet a new attribute constructed out of that attribute be assigned a high weight, due to a high correlation between the new constructed attribute and the goal one.

This approach can be regarded as a limited form of constructive induction, since the function used for constructing attributes (the product operator) is predetermined, rather than evolved. Only the pair of attributes being multiplied is evolved by the GA.

[Raymer et al. 1996] have extended this approach to evolve a combination of functions used for constructing new attributes. In essence, this extension was obtained by replacing the GA with a GP. Each GP individual consists of m trees, where m is the number of original attributes. Each of these trees, denoted t_i, encodes a new attribute constructed out of the i-th original attribute as a combination of functions applied to the value of that attribute. Each tree t_i is built by using the function set $\{+, -, *, /_{\text{protected}}\}$ and the terminal set $\{ERC, A_i\}$, where ERC is an ephemeral random constant and A_i is the original value of the i-th attribute. The basic idea of this individual encoding is illustrated in Figure 9.4. The GP is used as a wrapper around a k-nn algorithm. Hence, the k-nn algorithm is applied to the data modified by a GP individual, rather than to the original data. That is, each data instance's attribute A_i is passed through its corresponding tree t_i, and the output of these trees are used as the values of new constructed attributes, which are then given to the k-nn algorithm. The fitness function is based on the performance of the k-nn algorithm, following the spirit of the wrapper approach.

This form of individual encoding, with one tree for each original attribute (rather than one tree for all attributes), was chosen by Raymer et al. to better indicate the kind of function that each attribute requires in order to have higher predictive accuracy. This has the advantage of improving the understanding of the data. This advantage is obtained at the expense of ignoring attribute interactions during attribute construction, i.e., the GP misses relationships involving a combination of two or more attributes. In addition, if the number of attributes (m) is large then a GP individual would be very long, and its evaluation would be very time consuming. As pointed out by the authors, as m grows large this problem can be circumvented by encoding in an individual a single tree, representing a combination of all attributes.

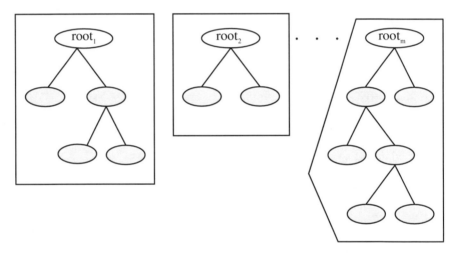

Figure 9.4: A GP individual representation with one tree for each new constructed attribute

9.3 Combining Attribute Selection and Attribute Weighting

A GA for simultaneous attribute selection and attribute weighting was proposed by [Raymer ct al. 1997] and later extended by [Raymer et al. 2000]. In this section we focus on the more recent version of this work, described in [Raymer et al. 2000]. The proposed GA works as a wrapper around an attribute-weighted k-nn (nearest neighbor) algorithm, and its basic ideas are described in the following.

An individual of the GA population consists of three substrings: the first one encodes a set of selected attributes, the second one encodes attribute weights, and the third one encodes some classifier-specific information. When using a k-nn algorithm as a classifier, the substring encoding classifier-specific information specifies the value of k, i.e., the number of nearest neighbors used for classification.

The attribute-weighting substring consists of a vector of m real-valued weights, where m is the number of attributes. Each position i of this vector – $i=1,...,m$ – contains the weight assigned to the i-th attribute.

The attribute-selection substring consists of a binary string indicating which attributes are selected. The standard individual encoding for attribute selection could be used here. As discussed in section 9.1.1.1, this encoding assigns a single bit – which we will call a selection bit – to each attribute. An attribute is considered selected if its corresponding selection bit is set to 1, whereas the attribute is considered removed if its corresponding selection bit is set to 0. However, the authors have instead chosen to assign several selection bits to each attribute. An attribute is considered selected only if the majority of its selection bits is set to 1. Their motivation for this approach was that single bits can cause discontinuities in the fitness function, making the problem more difficult for the GA.

The fitness function being minimized by the GA was the following:

$$Fitness = W_{errors}\#errors + W_{SelAttr}\#SelAttr + W_{votes}\#IncorrectVotes \\ + W_{diff\text{-}acc}Diff\_Acc, \quad where:$$

- $\#errors$ is the number of classification errors (incorrect predictions) in the training set;
- $\#SelAttr$ is the number of selected attributes;
- $\#IncorrectVotes$ is the number of incorrect votes during k-nn classification (this is equivalent to the *nmin* term mentioned in subsection 9.2.1);
- $Diff\_Acc$ is the difference in accuracy between classes;
- W_{errors}, $W_{SelAttr}$, W_{votes}, $W_{diff\text{-}acc}$ are user-defined weights.

In the above formula the *#errors* and *#SelAttr* terms have the obvious role of favoring classifications with a smaller number of errors and with a smaller attribute set. The term *#IncorrectVotes* is included to smooth the fitness function. When only the number of classification errors is used the fitness function tends to have a discrete stepwise character. This tends to occur because classification is done by using the class of the majority of the nearest neighbors, ignoring the vote of a minority of nearest neighbors. Hence, including the number of incorrect votes in the fitness function leads to a finer-grain, better-informed fitness function, considering the votes of all nearest neighbors. The *Diff_Acc* term is included to favor the correct prediction of all classes in a balanced way, rather than maximizing the number of correct predictions in some class at the expense of reducing the number of correct predictions in other classes.

9.4 GP for Attribute Construction

In the previous sections we saw that a simple GA can be naturally applied to some important data preparation tasks, namely attribute selection and attribute weighting. In those cases a candidate solution consisted simply of a subset of attributes or a set of attribute weights, respectively. In both cases a candidate solution consisted essentially of data, but not operations. Hence, in general an individual had a fixed length, and the overall shape of a candidate solution was fixed in advance. Only the contents of a candidate solution – the selected attributes or the attribute weights – were optimized by the GA.

In the case of attribute construction, however, the problem is in general considerably more difficult. As discussed in section 4.3, the general goal of an attribute construction algorithm is to construct new attributes by applying some operations/functions to the original attributes, such that the new attributes have more predictive power than the original attributes. This means an attribute construction algorithm has to solve at least two related problems. First, it has to determine which of the original attributes should be used to construct new attributes. Second, it has to determine which operations/functions are to be applied to the original attributes to construct new attributes.

In particular, this second problem intuitively suggests that Genetic Programming (GP) can be applied more naturally to this task than other "simpler" EAs, such as conventional genetic algorithms. After all, a GP individual explicitly represents a combination of data and operations/functions applied to that data.

In contrast, in most other kinds of EAs an individual explicitly represents only data, as mentioned above. These simpler kinds of evolutionary algorithms can still be used for constructing new attributes, but with some limitations. In general the operators/functions used for constructing new attributes are "implicit" in the genome-decoding procedure, rather than explicitly encoded in the genome. In other words, the operators/functions used for constructing new attributes are normally fixed, they are not subject to evolution. In general, only the combination of raw attributes used for constructing a new attribute is encoded in the genome and therefore evolved. An example of how to use a GA for constructive induction in this fashion is the work of [Punch et al. 1993], mentioned in subsection 9.2.2.

Although the use of GP in attribute construction seems promising, there is a caveat. The application of GP in attribute construction faces data typing problems similar to the ones discussed in section 7.1 in the context of rule discovery. In particular, recall that GP normally has to satisfy the property of closure – the output of a tree node should be able to be used as the input for any other tree node. This is a problem in attribute construction for the same basic reason that it is a problem in rule discovery: the data being mined normally contains attributes of different data types.

Fortunately, the solutions are similar to the ones discussed for rule discovery. One possible solution consists of modifying the data being mined such that all attributes are of the same data type and then use a function set containing only functions that receive input and produce output of that data type. For instance, one can booleanize all attributes and then use only boolean operators in the function set – see section 7.2. Alternatively, we can leave the data being mined as it is and use a constrained-syntax or strongly-typed GP – see section 7.3.

Each of the following two subsections presents an overview of a GP for attribute construction. In both cases the closure property is satisfied by converting all attributes to the same data type and then using functions that receive input and produce output of that data type. The GP described in subsection 9.4.1 follows the above-mentioned approach of booleanization, using a function set that contains only boolean functions; whereas the GP described in subsection 9.4.2 follows the approach of using a function set containing only general mathematical functions (+, -, *, /). This latter subsection also addresses the point that a GP using only general functions can be used to discover specialized, "problem-dependent" functions.

9.4.1 Constructing Combinations of Boolean Attributes

[Hu 1998] proposed a GP system called GPCI (Genetic Programming for Constructive Induction), which constructs attributes as a preprocessing step for a classification algorithm. In this system attributes are constructed in a way independent of the classification algorithm, so that the constructed attributes can be given as input to any classification algorithm.

The first step of GPCI is to booleanize attributes, i.e., to transform each continuous or categorical attribute into a boolean attribute. The basic idea is that each attribute A is transformed into two boolean attributes, denoted $Pos(A,v)$ and $Neg(A,v)$, partitioning all the values of the domain of A into two groups of values ("positive" and "negative" values), as follows:

For continuous (real-valued) attributes:
$Pos(A,v) = true$ if and only if the value of A is greater than v
$Pos(A,v) = false$ if and only if the value of A is less than or equal to v

For categorical (discrete) attributes:
$Pos(A,v) = true$ if and only if the value of A is equal to v
$Pos(A,v) = false$ if and only if the value of A is different from v

For each attribute A, the system finds the value v in its domain such that $Pos(A,v)$ has the highest ability to discriminate among examples of different classes in the training set. In CPGI this discrimination ability is evaluated by the well-known gain ratio measure [Quinlan 1993, chap. 2]. Once all attributes are booleanized, all pairs of $Pos(A,v)$ and $Neg(A,v)$ are available to be used as building blocks for constructing new attributes.

It should be noted that although the value v is chosen to "optimize" the class-discrimination ability of $Pos(A,v)$, this optimization is local – rather than global – in the sense that it is based only on the relationship between attribute A and the goal (or class) attribute. In other words, the interaction between attribute A and the other predictor attributes is ignored. However, some attribute interaction is taken into account in the next step of CPGI, which consists of running a GP for attribute construction.

The basic ideas of this GP are as follows. An individual represents a candidate new attribute. The terminal consists of all boolean attributes *Pos(A,v)* and *Neg(A,v)*. The function set contains boolean operators such as *AND* and *NOT*. A very simple example of an individual is shown in Figure 9.5, where *Age* is a continuous attribute and *Salary* is a categorical attribute. This individual represents the new attribute defined by: *(Age > 21)* AND *(Salary ≠ low)*.

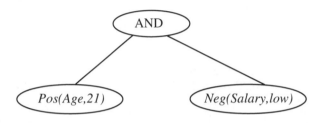

Figure 9.5: Very simple example of an individual for attribute construction

The fact that the new attributes are boolean combinations of original attributes not only simplifies the GP but also helps to keep the new attributes comprehensible for the user. This comprehensibility might be compromised if the new attributes were defined, say, by complex mathematical formulas combining numerical attributes. Of course, these advantages are obtained at the expense of missing other kinds of relationships between predictor attributes.

The fitness function of the GP is a measure of the quality of the new attribute associated with the individual. In CPGI the fitness function combines an absolute and a relative measure of attribute quality. The absolute measure directly evaluates the class-discrimination ability of the new attribute on training data, regardless of how the attribute was constructed. In contrast, the relative measure evaluates the new attribute's quality improvement relative to its "parent" attributes, i.e., the attributes from which the new attribute was constructed.

In CPGI the above-outlined GP is iteratively applied. Each GP run returns a new constructed attribute, and the training data is partitioned according to that attribute. This is conceptually similar to data partitioning according to a selected attribute in decision tree building (see subsection 3.1.1). The system is recursively called for each just-created data partition, until a stopping criteria is satisfied. For a more detailed description of the system, the reader is referred to [Hu 1998].

Hu compared CPGI with two other attribute construction methods, namely LFC [Ragavan et al. 1993] and GALA [Hu and Kibler 1996]. The comparison was made across 12 data sets. Overall, the predictive accuracy associated with GPCI was considerably better than LFC's one and somewhat better than GALA's one.

9.4.2 Discovering Specialized Functions

The conventional wisdom in GP is that we should include in the function set all the functions that are relevant for solving the target problem. In the context of attribute construction, this means we should include in the function set all functions that are useful for creating new attributes with greater predictive power than the original attributes. The problem, of course, is that we often do not know very well which functions are useful for a given application or a given set of original attributes.

In practice the correctness and completeness of our previous knowledge about the usefulness of functions for attribute construction is a matter of degree. For pedagogical purposes we can oversimplify the issue and divide the problem of constructive induction in two cases. In the first case we have a great deal of knowledge about which functions are useful for attribute construction, whereas in the second case we have little knowledge about this.

In the first case we can include in the function set only the relevant functions, which can be functions specialized for the data being mined. In this case the problem to be solved by the GP will be essentially to find which combinations of the relevant functions perform better for which combinations of original attributes. In the second case we often have to include in the function set some relatively general functions. In this case the problem to be solved by the GP is more difficult, and one can say that the system has more autonomy, in the sense that its performance is less dependent on the user providing a set of relevant functions. In some sense, the GP's task is not only to find which combinations of the general functions perform better for which combinations of original attributes, but also to discover more useful, specialized functions defined by combinations of the general functions included in the function set.

For instance, assume a relevant function for classification is the boolean eXclusive OR (XOR) function. (In general this will be typically the case in parity-like problems – see subsection 3.2.1.) If we knew this in advance we would include XOR in the function set. If not, we could include in the function set more general boolean functions, say AND and OR, rather than the more specific XOR function. In this case the GP should find a combination of AND and OR that corresponds to the definition of XOR.

Going further, still assuming that XOR is a relevant function, suppose now that we include in the function set only a very general set of mathematical functions, say $\{+, -, *, /_{protected}\}$. We still would like the GP to find a combination of these mathematical functions that corresponds to the definition of XOR. Does it sound unlikely? Well, it is not so unlikely as it seems at first glance.

Actually, Kuscu has shown that it is possible to automatically evolve mathematical expressions that correspond to logical functions [Kuscu 1996, 1999, 2000]. (One "trick" to map an expression composed of mathematical functions into an expression composed of boolean functions involves the use of a squashing function that maps the value of the original expression into a real number in the range 0...1. See [Kuscu 1996] for details.) The author even claims that [Kuscu 1999, p. 213]:

"...use of non-problem specific primitives such as arithmetical functions better promotes generalization for parity problems."

(Recall that parity problems are particularly difficult problems for rule discovery algorithms due to an extreme degree of attribute interaction, as discussed in subsection 3.2.1.)

In this work the GP is used as a wrapper around a classification algorithm, more precisely a backpropagation neural network algorithm. A GP individual encodes a predetermined number of constructed attributes. In the experiments reported in [Kuscu 1999] GP individuals encoded three new constructed attributes for parity problems. The function set contains only the mathematical functions $\{+, -, *, /_{\text{protected}}\}$ and the terminal set contains the original attributes. The constructed attributes are evaluated by adding them to the set of original attributes and giving all of these attributes to the neural network algorithm. The performance of the neural network algorithm is then fedback to the GP, in the form of a fitness function value for the GP individual.

Kuscu's experiments are interesting because they allow GP to have more autonomy than usual, increasing its potential for discovering new relevant functions for the target problem.

There is, however, one problem in using a function set having only arithmetic functions to evolve a logical function. The mathematical expressions evolved by GP will tend to be much longer (and so much less comprehensible) than their logical counterparts. One possible solution is to apply some kind of editing operator to the discovered rules, to make them shorter and more comprehensible.

9.5 Combining Attribute Selection and Construction with a Hybrid GA/GP

[Vafaie and DeJong 1998] proposed a hybrid GA/GP system where each individual represents a candidate solution for both attribute selection and attribute construction. The basic idea is that an individual is encoded as a string which contains two kinds of gene, namely a "simple gene" containing an original attribute or a "complex gene" containing a constructed attribute, formed by applying operations on a combination of original attributes. This hybrid encoding can be implemented in the style of the individual shown in Figure 9.6.

In this figure the attribute index A_i indicates that the i-th original attribute is selected. An original attribute is selected if it is present in one or more of the simple genes of the individual. Thus, the simple genes follow the basic idea of individual encoding for attribute selection discussed in subsection 9.1.1.2.

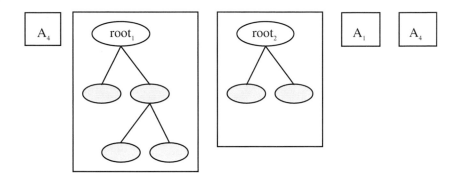

Figure 9.6: Example of individual in hybrid GA/GP for attribute selection/construction

On the other hand, constructed attributes are encoded in complex genes that follow the basic idea of GP's individual representation. Each constructed attribute is represented by a tree, where the leaf nodes are attribute indexes and the internal nodes are operations.

The system uses two forms of crossover, one at the attribute set level and the other at the attribute level. In the former crossover can occur only at gene boundaries, so that entire subsets of attributes (both selected and constructed) are swapped as a whole, without modifying the internal structure of a constructed attribute. In contrast, at the attribute level crossover is performed by swapping subtrees – in a conventional GP style – between a given pair of genes belonging to two parent individuals.

Mutation is also implemented at the attribute set level and attribute level. In the former the contents of a gene is simply replaced by one of the original attributes. At the attribute level a mutation of a constructed attribute is performed by randomly choosing a tree node to be modified and then replacing the current contents of that node with a new operator or original attribute, depending on whether the node being mutated is an internal or leaf node, respectively.

The evaluation of an individual follows the basic idea of the wrapper approach, i.e., the set of selected and constructed attributes contained in the individual is given as input to a classification algorithm, and the performance of this algorithm is directly used as a measure of fitness of the individual.

References

[Bala et al. 1995] J. Bala, K. De Jong, J. Huang, H. Vafaie and H. Wechsler. Hybrid learning using genetic algorithms and decision trees for pattern classification. *Proceedings of the International Joint Conference on Artificial Intelligence (IJCAI '95)*, 719–724. 1995.

[Bala et al. 1996] J. Bala, K. De Jong, J. Huang, H. Vafaie and H. Wechsler. Using learning to facilitate the evolution of features for recognizing visual concepts. *Evolutionary Computation 4(3)*, 297–312, 1996.

[Chen and Liu 1999] K. Chen and H. Liu. Towards an evolutionary algorithm: a comparison of two feature selection algorithms. *Proceedings of the Congress on Evolutionary Computation (CEC '99)*, 1309–1313. Washington D.C., 1999.

[Chen et al. 1999] S. Chen, C. Guerra-Salcedo, and S.F. Smith. Non-standard crossover for a standard representation – commonality-based feature subset selection. *Proceedings of the Genetic and Evolutionary Computation Conference (GECCO '99)*, 129–134. Morgan Kaufmann, 1999.

[Cherkauer and Shavlik 1996] K.J. Cherkauer and J.W. Shavlik. Growing simpler decision trees to facilitate knowledge discovery. *Proceedings of the 2nd International Conference on Knowledge Discovery and Data Mining (KDD '96)*, 315–318. AAAI Press, 1996.

[Emmanouilidis et al. 2000] C. Emmanouilidis, A. Hunter and J. MacIntyre. A multiobjective evolutionary setting for feature selection and a commonality-based crossover operator. *Proceedings of the 2000 Congress on Evolutionary Computation (CEC '2000)*, 309–316. IEEE, 2000.

[Freitas 1997] A.A. Freitas. The principle of transformation between efficiency and effectiveness: towards a fair evaluation of the cost-effectiveness of KDD techniques. *Proceedings of the 1st European Symposium on Principles of Data Mining and Knowledge Discovery (PKDD '97). Lecture Notes in Artificial Intelligence 1263*, 299–306. Springer, 1997.

[Freitas 2002] A.A. Freitas. A survey of evolutionary algorithms for data mining and knowledge discovery. To appear in: A. Ghosh and S. Tsutsui (Eds.) *Advances in Evolutionary Computation*. Springer, 2002.

[Goldberg 1989] D. Goldberg. *Genetic Algorithms in Search, Optimization and Machine Learning*. Addison-Wesley, 1989.

[Guerra-Salcedo and Whitley 1998] C. Guerra-Salcedo and D. Whitley. Genetic search for feature subset selection: a comparison between CHC and GENESIS. *Genetic Programming 1998: Proceedings of the 3rd Annual Conference*, 504–509. Morgan Kaufmann, 1998.

[Guerra-Salcedo and Whitley 1999a] C. Guerra-Salcedo and D. Whitley. Genetic approach to feature selection for ensemble creation. *Proceedings of the Genetic and Evolutionary Computation Conference (GECCO '99)*, 236–243. Morgan Kaufmann, 1999.

[Guerra-Salcedo and Whitley 1999b] C. Guerra-Salcedo and D. Whitley. Feature selection mechanisms for ensemble creation: a genetic search perspective. In: A.A. Freitas (Ed.) *Data Mining with Evolutionary Algorithms: Research Directions – Papers from the AAAI '99/GECCO '99 Workshop. Technical Report WS-99-06*, 13–17. AAAI Press, 1999.

[Guerra-Salcedo et al. 1999] C. Guerra-Salcedo, S. Chen, D. Whitley, and S. Smith. Fast and accurate feature selection using hybrid genetic strategies. *Proceedings of the Congress on Evolutionary Computation (CEC '99)*, 177–184. Washington D.C., USA. 1999.

[Hu 1998] Y-J. Hu. A genetic programming approach to constructive induction. *Genetic Programming 1998: Proceedings of the 3rd Annual Conference,* 146–151. Morgan Kaufmann, 1998.

[Hu and Kibler 1996] Y-J. Hu and D. Kibler. Generation of attributes for learning algorithms. *Proceedings of the 1996 National Conference on Artificial Intelligence (AAAI '96),* 806–811. AAAI Press, 1996.

[Ishibuchi and Nakashima 2000] H. Ishibuchi and T. Nakashima. Multi-objective pattern and feature selection by a genetic algorithm. *Proceedings of the 2000 Genetic and Evolutionary Computation Conference (GECCO '2000),* 1069–1076. Morgan Kaufmann, 2000.

[Kim et al. 2000] Y. Kim, W.N. Street and F. Menczer. Feature selection in unsupervised learning via evolutionary search. *Proceedings of the 6th ACM SIGKDD International Conference on Knowledge Discovery and Data Mining (KDD '2000),* 365–369. ACM, 2000.

[Kudo and Skalansky 2000] M. Kudo and J. Sklansky. Comparison of algorithms that select features for pattern classifiers. *Pattern Recognition 33(2000),* 25–41.

[Kuscu 1996] I. Kuscu. Evolution of learning rules for hard learning problems. *Proceedings of the 5th Annual Evolutionary Programming Conference* MIT Press, 1996.

[Kuscu 1999] I. Kuscu. A genetic constructive induction model. *Proceedings of the Congress on Evolutionary Computation (CEC '99),* 212–217. Washington D.C., 1999.

[Kuscu 2000] I. Kuscu. Generalisation and domain specific functions in genetic programming. *Proceedings of the Congress on Evolutionary Computation (CEC '2000),* 1393–1400. IEEE, 2000.

[Llora and Garrel 2001] X. Llora and J.M. Garrell. Inducing partially-defined instances with evolutionary algorithms. *Proceedings of the 18th International Conference on Machine Learning (ICML '2001),* 337–344. Morgan Kaufmann, 2001.

[Martin-Bautista and Vila 1999] M.J. Martin-Bautista and M.-A. Vila. A survey of genetic feature selection in mining issues. *Proceedings of the Congress on Evolutionary Computation (CEC '99),* 1314–1321. IEEE, 1999.

[Menczer et al. 2000] F. Menczer, M. Degeratu and W.N. Street. Efficient and scalable Pareto optimization by evolutionary local selection algorithms. *Evolutionary Computation 8(2),* 223–247, 2000.

[Moser and Murty 2000] A. Moser and M.N. Murty. On the scalability of genetic algorithms to very large-scale feature selection. *Proceedings of the Real-World Applications of Evolutionary Computing (EvoWorkshops 2000). Lecture Notes in Computer Science 1803,* 77–86. Springer, 2000.

[Punch et al. 1993] Punch, W.F.; Goodman, E.D.; Pei, M.; Chia-Shun, L.; Hovland, P. and Enbody, R. Further research on feature selection and classification using genetic algorithms. *Proceedings of the 5th International Conference Genetic Algorithms (ICGA '93),* 557–564.

[Quinlan 1993] J.R. Quinlan. *C4.5: Programs for Machine Learning.* Morgan Kaufmann, San Mateo, 1993.

[Ragavan et al. 1993] H. Ragavan, L. Rendell, M. Shaw and A. Tessmer. Complex concept acquisition through direct search and feature caching. *Proceedings of the 13th International Joint Conference on Artificial Intelligence (IJCAI '93)*, 946–951. 1993.

[Raymer et al. 1996] M.L. Raymer, W.F. Punch, E.D. Goodman and L.A. Kuhn. Genetic programming for improved data mining - application to the biochemistry of protein interactions. *Genetic Programming 1996: Proceedings of the 1st Annual Conference,* 375–380. Morgan Kaufmann, 1996.

[Raymer et al. 1997] M.L. Raymer, W.F. Punch, E.D. Goodman, P.C. Sanschagrin and L.A. Kuhn. Simultaneous feature scaling and selection using a genetic algorithm. *Proceedings of the 7th International Conference on Genetic Algorithms (ICGA '97)*, 561–567. Morgan Kaufmann, 1997.

[Raymer et al. 2000] M.L. Raymer, W.F. Punch, E.D. Goodman, L.A. Kuhn and A.K. Jain. Dimensionality reduction using genetic algorithms. *IEEE Transactions on Evolutionary Computation 4(2)*, 164–171, 2000.

[Sharpe and Glover 1999] P.K. Sharpe and R.P. Glover. Efficient GA based techniques for classification. *Applied Intelligence 11*, 277–284, 1999.

[Sherrah et al. 1997] J.R. Sherrah, R.E. Bogner and A. Bouzerdoum. The evolutionary pre-processor: automatic feature extraction for supervised classification using genetic programming. *Genetic Programming 1997: Proceedings of the 2nd Annual Conference (GP '97)*, 304–312. Morgan Kaufmann, 1997.

[Terano and Ishino 1998] T. Terano and Y. Ishino. Interactive genetic algorithm based feature selection and its application to marketing data analysis. In: H. Liu and H. Motoda (Eds.) *Feature Extraction, Construction and Selection*, 393–406. Kluwer, 1998.

[Thompson 1998] S. Thompson. Pruning boosted classifiers with a real valued genetic algorithm. *Research and Develop. in Expert Systems XV – Proceedings of ES'98*, 133–146. Springer, 1998.

[Thompson 1999] S. Thompson. Genetic algorithms as postprocessors for data mining. In: A.A. Freitas (Ed.) *Data Mining with Evolutionary Algorithms: Research Directions – Papers from the AAAI Workshop*, 18–22. Technical Report WS-99-06. AAAI Press, 1999.

[Vafaie and DeJong 1998] H. Vafaie and K. DeJong. Evolutionary Feature Space Transformation. In: H. Liu and H. Motoda (Eds.) *Feature Extraction, Construction and Selection*, 307–323. Kluwer, 1998.

[Wang and Sundaresh 1998] K. Wang and S. Sundaresh. Selecting features by vertical compactness of data. In: H. Liu and H. Motoda (Eds.) *Feature Extraction, Construction and Selection*, 71–84. Kluwer, 1998.

[Yang and Honavar 1997] J. Yang and V. Honavar. Feature subset selection using a genetic algorithm. *Genetic Programming 1997: Proceedings of the 2nd Annual Conference (GP '97)*, 380–385. Morgan Kaufmann, 1997.

[Yang and Honavar 1998] J. Yang and V. Honavar. Feature subset selection using a genetic algorithm. In: H. Liu and H. Motoda (Eds.) *Feature Extraction, Construction and Selection*, 117–136. Kluwer, 1998.

10 Evolutionary Algorithms for Discovering Fuzzy Rules

This chapter discusses several concepts and issues in the development of Evolutionary Algorithms (EAs) for discovering fuzzy prediction rules. We start with a review of basic concepts of fuzzy sets, in section 10.1, and with a discussion on the difference between fuzzy prediction rules and crisp prediction rules, in section 10.2.

Then section 10.3 presents a simple taxonomy of EAs for fuzzy-rule discovery. Section 10.4 discusses how to use EAs only for generating fuzzy rules. Section 10.5 discusses how to use EAs only for tuning membership functions. Finally, section 10.6 discusses a more complex scenario where EAs are used for both generating fuzzy rules and tuning membership functions.

We emphasize that this chapter is not about fuzzy EAs. Rather, this chapter is about discovering fuzzy prediction rules with EAs. In general, unless mentioned otherwise, we will assume that the genetic operators used by the EA, as well as the algorithm's parameters, are crisp. In order to discover fuzzy rules, usually only the individual representation and the fitness function need to be fuzzified, as will be seen later.

10.1 Basic Concepts of Fuzzy Sets

10.1.1 Fuzzy Sets

A fuzzy set is a set whose boundaries are not clearly defined. An object belongs to a fuzzy set to a certain degree, called the degree of membership, typically represented by a real-valued number in the interval [0..1]. This is in contrast with conventional, crisp sets, whose boundaries are clearly defined. An object either belongs or does not belong to a crisp set. In terms of degrees of membership, one can say that an object can belong to a crisp set with only one out of two possible degrees of membership: 1 or 0.

As a simple example, consider two alternative definitions of the set of *old* people. A crisp definition could be: a person belongs to the set of *old* people if and only if her $Age \geq 65$ years. This definition is illustrated in Figure 10.1(a). The problem with this definition is that there is a very abrupt change in the status of a person, from *not old* to *old*. In the day of her 65-year birthday the person status abruptly changes from *not old* to *old*, which clearly is not a reasonable description of reality.

In contrast, a fuzzy definition would recognize that people gradually get older. A possible definition for the fuzzy set *old* is illustrated in Figure 10.1(b). In this example a person whose *Age* is less than 60 years is considered *old* to a degree of 0. From that *Age* onwards the degree of *oldness* of a person gradually increases, until the person reaches the *Age* of 70 years. From that *Age* onwards the person is considered *old* to a degree of 1. This is a more reasonable description of reality than the crisp description of Figure 10.1(a).

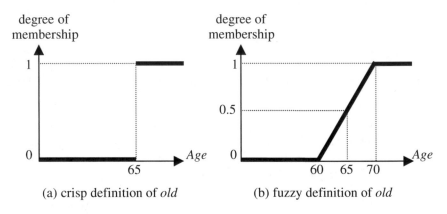

(a) crisp definition of *old*   (b) fuzzy definition of *old*

Figure 10.1: Crisp versus fuzzy definition of the set of *old* people

It should be noted that the concept of fuzzy sets can be considered a generalization of the concept of crisp sets, since in the former degrees of membership can take on any value in the continuous interval [0..1], whereas in the latter degrees of membership can take on only the value 0 or 1.

Figure 10.1(b) illustrated how one can fuzzify an attribute value such as *old*. In practice one fuzzifies all the values of an attribute. This is illustrated in Figure 10.2. In this figure the attribute *Age* is fuzzified into three linguistic values, namely *young*, *middle-aged*, and *old*. Each of these linguistic values is associated with a fuzzy set, which is defined by its membership function.

Note that in Figure 10.2 some fuzzy sets overlap, so that a person with a given value of *Age* can belong to more than one fuzzy set. For instance, a person with an *Age* of 63 belongs to the set of *middle-aged* people to a degree of 0.7, to the set of *old* people to a degree of 0.3, and to the set of *young* people to a degree of zero. In this example the summation of these degrees of membership (0.7 + 0.3 + 0) equals 1. This property holds for all values of *Age* in Figure 10.2. This restriction is intuitive and will be often used in this chapter, even though this is not a necessary requirement of fuzzy sets.

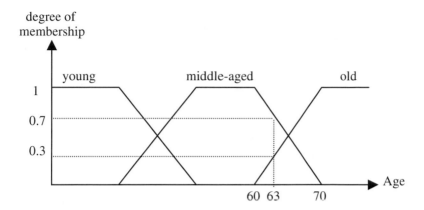

Figure 10.2: Attribute *Age* being fuzzified into three linguistic values

The cardinality of a crisp set is the number of elements belonging to that set. In the case of a fuzzy set A, each element x belongs to A to a certain degree, given by its membership value $\mu_A(x)$. Thus, given a finite universal set X, generalizing the above definition of cardinality one can define the cardinality of a fuzzy set A, denoted by $|A|$, as:

$$|A| = \sum_{x \in X} \mu_A(x) .$$

For instance, suppose that X contains just five elements x_1, x_2, x_3, x_4, x_5, which belong to fuzzy set A to degrees of 0, 0.2, 0.7, 0.9, 1. Then $|A| = 2.8$.

10.1.2 Operations on Fuzzy Sets

Operations on fuzzy sets are generalizations of operations on crisp sets. This generalization can be done in several different ways. As a result, each operation on a fuzzy set – such as fuzzy complement, fuzzy intersection and fuzzy union – can be defined in many different ways. We first review the "standard" definition of fuzzy operations, and later on we will review a couple of alternative definitions.

In the following we use the notation $\mu_A(x)$ to denote a membership function that maps all elements x of a given universal set X into a real number in [0..1], representing the degree to which x belongs to the fuzzy set A. Note that the universal set X is a crisp set. For the purposes of this chapter, the universal set corresponds to the set of values of an attribute of the data being mined. We will refer to such an attribute as an attribute being fuzzified. Hence, the elements x are all possible values of an attribute being fuzzified. For instance, consider the fuzzy set *old* defined by the corresponding membership function shown in Figure 10.2. Then, for the *Age* value $x = 63$ we have $\mu_{old}(63) = 0.3$.

To compute the standard complement of a fuzzy set A, denoted *NOT A*, we have to determine the degree to which each element x belongs to *NOT A*. This is defined as:

$$\mu_{NOT\text{-}A}(x) = 1 - \mu_A(x).$$

For instance, given the membership function for the linguistic value *old* shown in Figure 10.2, $\mu_{NOT\text{-}old}(63) = 1 - \mu_{old}(63) = 1 - 0.3 = 0.7$.

To compute the standard intersection of two fuzzy sets A and B, denoted $A \cap B$, we have to determine the degree to which each element x belongs to $A \cap B$. This is defined as:

$$\mu_{A\cap B}(x) = \min[\mu_A(x), \mu_B(x)] \,,$$

where min denotes the minimum operator. For instance, given the membership functions for the linguistic values *old* and *middle-aged* shown in Figure 10.2, $\mu_{old\cap middle\text{-}aged}(63) = \min[\mu_{old}(63), \mu_{middle\text{-}aged}(63)] = \min[0.3, 0.7] = 0.3$.

To compute the standard union of two fuzzy sets A and B, denoted $A \cup B$, we have to determine the degree to which each element x belongs to $A \cup B$. This is defined as:

$$\mu_{A\cup B}(x) = \max[\mu_A(x), \mu_B(x)] \,,$$

where max denotes the maximum operator. For instance, given the membership functions for the linguistic values *old* and *middle-aged* shown in Figure 10.2, $\mu_{old\cup middle\text{-}aged}(63) = \max[\mu_{old}(63), \mu_{middle\text{-}aged}(63)] = \max[0.3, 0.7] = 0.7$.

As mentioned earlier, the above standard operations on fuzzy sets are not the only possible generalizations of their crisp counterpart. Many alternative definitions of operations on fuzzy sets are possible. The effectiveness of a given kind of operation, like the effectiveness of a given kind of membership function, is context-dependent, i.e., it depends on the application domain. However, there are some basic properties that are usually required from any operation on fuzzy sets, to make it reasonably intuitive. These basic properties are usually posed as axioms for defining operations on fuzzy sets. More precisely, fuzzy intersection and fuzzy union operations are defined as function mappings from $[0,1] \times [0,1]$ to $[0,1]$ – from a pair of membership degrees to another membership degree – that satisfy a given set of axioms [Klir and Yuan 1995; Yen 1999].

Table 10.1 summarizes the axioms normally used for defining fuzzy intersection and fuzzy union operations. In this table the fuzzy intersection $\mu_{A\cap B}(x)$ and the fuzzy union $\mu_{A\cup B}(x)$ are denoted by $i(a,b)$ and $u(a,b)$, respectively, where a and b are a shorter notation for $\mu_A(x)$ and $\mu_B(x)$.

Table 10.2 summarizes some definitions of fuzzy intersection and fuzzy union operations that satisfy the axioms of Table 10.1. Table 10.2 uses the same notation as Table 10.1. A detailed discussion about the operations shown in Table 10.2 is beyond the scope of this chapter, and it can be found in fuzzy systems textbooks, such as [Klir and Yuan 1995].

Table 10.1: Axioms of fuzzy intersection and fuzzy union

Axiom	Fuzzy intersection	Fuzzy union
A.1 (boundary condition)	$i(a,1) = a$ $i(a,0) = 0$	$u(a,0) = a$ $u(a,1) = 1$
A.2 (monotonicity)	$b \le c$ implies $i(a,b) \le i(a,c)$	$b \le c$ implies $u(a,b) \le u(a,c)$
A.3 (commutativity)	$i(a,b) = i(b,a)$	$u(a,b) = u(b,a)$
A.4 (associativity)	$i(a,i(b,c)) = i(i(a,b),c)$	$u(a,u(b,c)) = u(u(a,b),c)$

Here we just mention an important characteristic of the standard fuzzy intersection and fuzzy union operations. The standard fuzzy intersection produces for any given fuzzy sets the largest fuzzy set from among those produced by all fuzzy intersections that satisfy the axioms of Table 10.1. In other words, for any element x that belongs to fuzzy sets A and B with degrees a, $b \in [0,1]$, the value of the standard intersection $\mu_{A \cap B}(x) = \min(a,b)$ will be greater than or equal to the value of any other fuzzy intersection satisfying the axioms of Table 10.1. For instance, let $a = 0.2$ and $b = 0.9$. Then the values of standard, product and bounded difference intersections are respectively 0.2, 0.18 and 0.1.

Conversely, the standard fuzzy union produces the smallest fuzzy set among the fuzzy sets produced by all fuzzy unions that satisfy the axioms of Table 10.1. In other words, for any element x that belongs to fuzzy sets A and B with degrees a, $b \in [0,1]$, the value of the standard union $\mu_{A \cup B}(x) = \max(a,b)$ will be smaller than or equal to the value of any other fuzzy union satisfying the axioms of Table 10.1. For instance, let $a = 0.2$ and $b = 0.9$. Then the values of standard, product and bounded difference unions are respectively 0.9, 0.92 and 1.

Table 10.2: Some definitions of operations on fuzzy sets

intersection	
standard	$\min(a,b)$
algebraic product	$a \times b$
bounded difference	$\max(0, a + b - 1)$

union	
standard	$\max(a,b)$
algebraic sum	$a + b - a \times b$
bounded sum	$\min(1, a + b)$

10.1.3 Membership Functions

Membership functions can be defined in a number of different ways, based on different shapes and different number of parameters. Figures 10.3 through 10.8 illustrate some possible definitions. Figure 10.3 shows a trapezoidal membership function defined by four parameters, namely the coordinates of points p_1, p_2, p_3 and p_4. Of course, fewer parameters would be enough if we imposed some restrictions on the trapezoidal shape. For instance, if we want only "symmetric" trapezoids, where the distance between p_1 and p_2 is equal to the distance between

p_3 and p_4, then we could replace parameters p_1 and p_4 with a single parameter specifying this distance.

Figure 10.4 shows a triangular membership function defined by two parameters, namely the coordinate of a point p, for which the membership function has its maximum value, and the width of the base of the triangle. This figure assumes that the triangle is isosceles, so that point p divides the triangle's base into two equal-length segments.

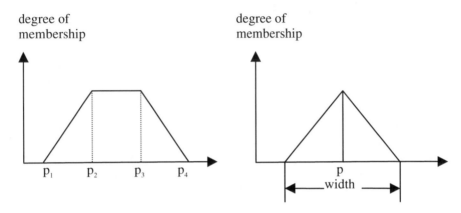

degree of membership

degree of membership

Figure 10.3: trapezoidal memb. funct.   **Figure 10.4:** triangular memb. funct.

Of course, different shapes are suitable for different applications. Suppose we want to define a membership function to represent the linguistic value *middle-aged* of the attribute *Age*. Consider, for instance, *Age* values in the interval 40...45. It seems reasonable that all *Age* values in this interval should have the maximal degree of membership (100%) in the set of *middle-aged* people. It can be argued that in this case trapezoidal functions are more intuitive than triangular functions. After all, in the former the maximal value of membership is associated with an interval of attribute values, represented by the coordinates of the top edge of the trapezoid. In triangular functions, on the other hand, the maximal value of membership is associated with a single value of the attribute being fuzzified, represented by the coordinate of the single point at the top of the triangle.

Now suppose that, in another application, we want to define a membership function to represent the linguistic value *close-to-21-years* of the attribute *Age*. In this case a triangular membership function seems more intuitive than a trapezoidal one. In the former we can naturally associate the *Age* value of 21 with the point p shown in Figure 10.4; whereas in the latter the choice of coordinates p_2 and p_3 in Figure 10.3 would be somewhat arbitrary.

Note that both Figures 10.3 and 10.4 show the definition of a single membership function, corresponding to a single linguistic value. In practice an attribute is fuzzified into two or more linguistic values and it is desirable to make sure that the corresponding membership functions interact in a meaningful way. For instance, Figure 10.5 shows a situation where the interaction between three membership functions would be counterintuitive in most applications. This interaction has at least two drawbacks. First, there is a range R of values of X, namely between p_1 and p_2, where all the elements $x \in R$ do not belong to any set – or, equivalently, belong to a zero degree to all sets. Second, there is not much distinction between the linguistic values *medium* and *large*, which tends to be counterintuitive.

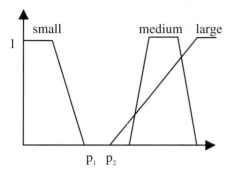
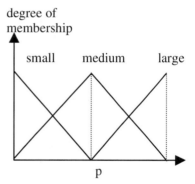

Figure 10.5: counterintuitive membership functions

Figure 10.6: very coarse-grained fuzzification

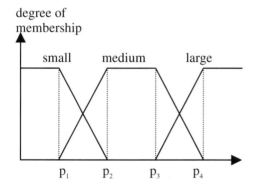

Figure 10.7: 3 ling. values represented by trapezoidal memb. funct.

Figure 10.8: 3 ling. values represented by triangular memb. funct.

Sometimes, although the interaction between membership functions fuzzifying an attribute is specified in a meaningful way, the degree of granularity of the fuzzification is not suitable for the target application. For instance, in Figure 10.6 the fuzzification of the universal set X is very coarse-grained, since X is fuzzified by using only two linguistic values. Suppose that X is the set of values that can be taken on by the attribute *Age*. Intuitively, in most applications this fuzzification would be too coarse-grained, since describing, say, the *Age* value of 40 as *young* or *old* would be quite counterintuitive.

Figure 10.7 illustrates a fuzzification of the universal set X into three linguistic values, each of them defined by a trapezoidal membership function. In this figure the interaction between membership functions seems to be more sensible than the one in Figure 10.5, for most applications. Note that in Figure 10.7 three trapezoidal membership functions are defined by the same set of four parameters p_i – i=1,...,4 – used to define a single membership function in Figure 10.3. The need for a larger number of parameters in Figure 10.7 was avoided by imposing some restrictions on the membership functions. In particular, each point p_i corresponds to an inflexion point for two membership functions. For instance, p_1 specifies both the point where the value of the membership function *medium* starts to rise above zero and the point where the value of the membership function *small* starts to fall below 1.

Figure 10.8 illustrates a fuzzification where three triangular membership functions are defined by a single parameter p. This minimization of the number of parameters was achieved by imposing strong restrictions on the three membership functions, as can be seen in the figure.

Note that Figures 10.3 through 10.8 have shown only piecewise-linear membership functions, where each piece of membership function can be described by a linear equation. For instance, let the values of p_1, p_2, p_3 and p_4 in Figure 10.7 be 10, 20, 30 and 40, respectively. The membership function *medium* in this figure can be described in a piecewise fashion by the following equations:

$\mu_{medium}(x) = 0$ if $x \le 10$,
$\mu_{medium}(x) = 0.1(x - 10) = 0.1x - 1$ if $10 < x \le 20$,
$\mu_{medium}(x) = 1$ if $20 < x \le 30$,
$\mu_{medium}(x) = -0.1(x - 40) = -0.1x + 4$ if $30 < x \le 40$,
$\mu_{medium}(x) = 0$ if $x > 40$.

Piecewise-linear membership functions are simple and fast to compute. However, in some applications non-linear membership functions may be more appropriate. For instance, [Cox 1995, p. 229] claims that a kind of Gaussian membership function "...*offers a better semantic decomposition for most real-world business problems.*" [Crockett et al. 2000] discuss several non-linear membership functions in the context of fuzzy decision trees.

10.1.3.1 *Guidelines for Defining Membership Functions*

As mentioned above, the kind of membership function to be used depends on the application. In any case, some general guidelines may be useful to design "sensible" membership functions. For instance, [Hirota and Pedrycz 1999; Espinosa and Vandewalle 2000; Setnes and Roubos 2000] suggest very similar guidelines to define fuzzy sets that are "semantically sound", or have a good "linguistic interpretability", as follows:

(a)   Each fuzzy set should be unimodal and normal. (In essence, a fuzzy set is normal when the largest membership degree obtained by any element in the set is 1.)

(b)   The fuzzy sets (or, equivalently, the corresponding membership functions) used to fuzzify an attribute should be sufficiently disjoint. This ensures that the linguistic values associated with the fuzzy sets are sufficiently distinct, so that they are linguistically meaningful. For instance, as mentioned above, the fuzzy sets *medium* and *large* shown in Figure 10.5 fail to meet this condition, since they are not sufficiently disjoint.

(c)   The number of fuzzy sets (or linguistic values) used to fuzzify an attribute should be relatively low, say at most about seven. This number is based on some psychological research on limits of human cognition.

(d)   Any element of the universe of discourse (any original value of the attribute being fuzzified) should belong to at least one of the fuzzy sets used to fuzzify the attribute.

In addition, minimizing the number of parameters defining the membership functions is important if there are many attributes to be fuzzified and the values of the parameters are to be automatically optimized by a relatively slow search method, such as an EA or a neural network. In these cases restrictions imposed on the membership functions, such as the restrictions discussed above for Figures 10.7 and 10.8, help to significantly reduce the search space.

[Janikow 1995] proposed several kinds of restrictions that can be used to reduce the size of the search space considered by an algorithm that tries to optimize membership functions. This work used an EA for optimizing membership functions to be used in a fuzzy decision tree algorithm, but of course the proposed restrictions can be used by other kinds of algorithms.

10.2 Fuzzy Prediction Rules vs Crisp Prediction Rules

To understand the basic difference between crisp and fuzzy rules, let us consider two simple prediction rule antecedents, one of them crisp and the other one a fuzzified version of the former, as follows:

crisp rule: IF *(Age* ≤ *25)*   AND *(Sex* = *male)* THEN . . .
fuzzy rule: IF *(Age* is *young)* AND *(Sex* = *male)* THEN . . .

Suppose that the system has to compute the matching between each of these rules and a data instance with the value *23* for the attribute *Age* and the value *male* for the attribute *Sex*. In the case of the crisp rule, one can easily see that the data instance satisfies both rule conditions – it matches the rule to a degree of 100%.

In the case of the fuzzy rule the system has to use a membership function to compute the matching between the *Age* value of 23 and the linguistic value "*young*". Assume the attribute *Age* is fuzzified by the three membership functions shown in Figure 10.9. Then the system finds that the data instance satisfies the fuzzy condition *(Age* is *young)* to a degree of 0.7, and it satisfies the crisp condition *(Sex* = *male)* to a degree of 1. Next the system has to determine the degree to which the data instance satisfies the rule antecedent [(*Age* is *young*) AND (*Sex* = *male*)] as a whole. This requires the application of a fuzzy AND operator. Assuming the use of the standard fuzzy intersection (see section 10.1.2), the degree to which the data instance satisfies the rule antecedent as a whole would be given by min(0.7 , 1) = 0.7. The computation of the degree of matching between a fuzzy rule and a data instance will be discussed in somewhat more detail in subsection 10.4.2.

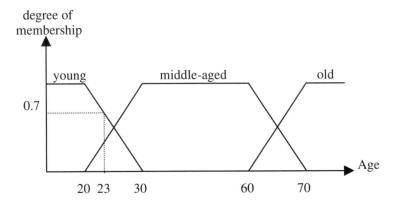

Figure 10.9: Attribute *Age* being fuzzified into three linguistic values

The above simple example also illustrates that in general only continuous attributes (e.g., *Age*) are fuzzified, i.e., categorical attributes (e.g., *Sex*) are kept crisp. Sometimes, however, one can fuzzify a categorical attribute whose values can be ordered. Consider, e.g., the attribute *Highest_Academic_Qualification*, which can take on a value in the domain {*primary_school, high_school, B.Sc.(or B.A.), M.Sc.(or M.A.), Ph.D.*}. It is possible – though not necessarily desirable – to fuzzify this attribute by using, say, two linguistic values: *low* and *high*.

As a final remark before we leave this section, we must draw attention to the fact that in the above example we have considered the occurrence of fuzzy conditions in the rule antecedent, but not in the rule consequent (the THEN part of the rule). Indeed, throughout this chapter we will assume that the rule consequent is crisp.

In theory one could fuzzify the rule consequent as well. For instance, suppose we want to discover a classification rule predicting whether the credit of a person is *good* or *bad*, to decide whether or not we grant the person a credit card. It could be argued that it would be natural to discover a rule with a fuzzy consequent, i.e., a rule where the goal attribute value *good* or *bad* was a linguistic value. Unfortunately, such rule would be less useful for decision making in our crisp, artificial financial society than a rule with a crisp consequent. At the end of the day we want to classify a person's credit into "*good*" or "*bad*" in a crisp way, since we will either grant the person a credit card or not. There is no way to grant the person "half a credit card". Hence, at least while our artificial society keeps requiring crisp decisions, crisp rule consequents tend to be more useful and more "natural", in practice, than fuzzy rule consequents.

10.3 A Simple Taxonomy of EAs for Fuzzy-Rule Discovery

In the next sections of this chapter we discuss how to adapt EAs for discovering fuzzy rules. There are at least three different approaches for performing this adaptation, as shown in the simple taxonomy of Figure 10.10. Each of these approaches is discussed separately in subsections 10.4, 10.5 and 10.6, respectively.

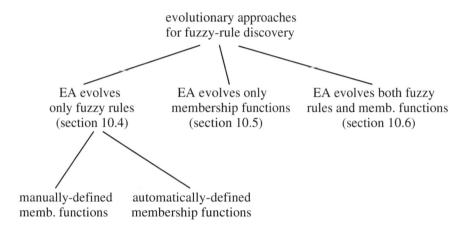

Figure 10.10: Some ways of using an EA for discovering fuzzy rules

10.4 Using EAs for Generating Fuzzy Rules

In this approach an EA is used to search for a good set of fuzzy prediction rules, but the membership functions of attribute values are assumed to be predefined, i.e., they are not evolved by the EA. (The issue of how to define good membership functions will be discussed later.) Therefore, the EA is used to find combinations of attribute values that are relevant for accurately predicting the value of another attribute.

Note that this is conceptually similar to the use of EAs for discovering crisp prediction rules. The main difference is that in the latter case all rules conditions are crisp, while in the former case the rule can have fuzzy rule conditions, as discussed in section 10.2.

We emphasize that in both the fuzzy-rule and the crisp-rule cases the EA is essentially searching for good combinations of attribute values. Therefore, many aspects of an EA designed for the latter case can be re-used in the former case. In particular, the methods of encoding crisp rules into individuals discussed in section 6.1 can be used, virtually without any modification, for encoding fuzzy rules.

Actually, from the viewpoint of rule encoding, a fuzzy rule condition such as *Age = young*, where *young* is a linguistic value associated with a fuzzy set, is not different from a crisp rule condition involving a categorical, discrete attribute such as *Age = 0...25*, where the attribute *Age* was previously discretized into crisp intervals such as *0...25*. In both cases we have a rule condition of the form $Attr_i = Val_{ij}$, where Val_{ij} is the *j*-th value belonging to the domain of the *i*-th attribute.

A similar argument can be used for most genetic operators. Fragments of the genotype corresponding to fuzzy attribute values, encoded either in a binary representation or in a higher-level representation, can be crossed over or mutated in the same way that the EA would cross over or mutate crisp, categorical attribute values.

However, the design of an EA that discovers fuzzy rules has an additional component, in comparison with EAs discovering crisp rules. This additional component involves the fact that in the fuzzy case the matching between a data instance and a rule is somewhat more complex.

Summarizing, one can say that, from a genotypic viewpoint (concerning individual encoding and genetic operators), an EA for discovering fuzzy rules can be designed in almost the same way as an EA for discovering crisp rules. However, from a phenotypic viewpoint (concerning individual decoding and fitness function computation) there are profound differences between the two kinds of EA, as will be seen in the next subsections.

The rest of this section is organized as follows. Subsection 10.4.1 discusses the individual encoding of fuzzy rules. Subsection 10.4.2 addresses the issue of determining the degree of matching between a rule antecedent and a data instance. Subsection 10.4.3 discusses the use of a set of fuzzy rules for classifying a given data instance. Subsection 10.4.4 discusses how membership functions can be specified in a way separate from the EA that is evolving the fuzzy rules.

10.4.1 Individual Encoding

In this subsection we first discuss the issue of encoding a fuzzy rule antecedent (the IF part of the rule) into an individual, and then we discuss the issue of determining the corresponding rule consequent (the THEN part of the rule).

10.4.1.1 Encoding a Fuzzy Rule Antecedent into an Individual

As mentioned above, the encoding of a fuzzy rule, or a set of fuzzy rules, into an individual involves some choices that are similar to the encoding of crisp rules into individuals. For instance, in the case of fuzzy rules we also have to make a choice between the following options:

Pittsburgh vs. Michigan Approach – Recall that in the Pittsburgh approach each individual of the EA population represents a set of prediction rules, i.e., an entire candidate solution; whereas in the Michigan approach each individual represents a single rule, i.e., a part of a candidate solution (a rule set).

Encoding Rule Conditions: binary vs. high-level encoding – When using a binary, low-level encoding, in general each predictor attribute $Attr_i$ will be assigned a certain number of bits, which depends on the data type of the attribute. Instead of using a binary encoding, we can encode attribute values directly in the genotype. This approach seems particularly useful in the case of continuous attributes, where a binary encoding tends to be somewhat cumbersome.

Fixed-Length vs. Variable-Length Genotype – Different prediction rules can have, of course, different numbers of conditions in their antecedent. There are two basic approaches to encode a variable-length rule antecedent. First, we can use a fixed-length genotype which is suitably decoded into a variable-length rule antecedent. Second, we can use a variable-length genotype that is directly equivalent to a variable-length rule antecedent, so that the number of conditions of a rule antecedent can increase or decrease during the evolution of the individuals.

The general pros and cons of each of the above options were discussed in detail in section 6.1, in the context of crisp rules. In the following we focus only on explaining the main differences in individual encoding associated with fuzzy rules. To simplify our discussion we will assume that the following individual-encoding decisions have already been made: (a) The EA follows the Michigan approach, i.e., each individual represent a single rule. (b) The EA uses a high-level encoding for rule conditions. A variable-length rule is encoded into an individual as a fixed-length genotype, where some genes (rule conditions) can take on the "don't care" value to indicate that the condition is effectively removed from the rule.

Figure 10.11 illustrates the use basic idea of an individual encoding in this case. In the Figure a rule antecedent is encoded into an individual as a set of m conditions, where m is the number of predictor attributes. Each condition $cond_i – i = 1,...,m$ – is in turn encoded as a quadruple: $<Attr_i, Op_i, Val_{ij}, Active_i>$, where:

- $Attr_i$ denotes the i-th predictor attribute;
- Op_i denotes a comparison operator;
- Val_{ij} denotes the j-th value of the domain of $Attr_i$;
- $Active_i$ is a bit used as a flag to indicate whether the i-th condition is active ("1") or inactive ("0"). By active we mean the condition is included in decoded rule; whereas by inactive we mean the condition is not included in the decoded rule.

Figure 10.11 is actually the same as Figure 6.2. The main difference is in the values that can be taken by Op_i and Val_{ij}, $i=1,...,m$. In the context of crisp rules Val_{ij} could take on either a nominal (discrete) or continuous value, depending on whether attribute $Attr_i$ is categorical or continuous. The comparison operator Op_i would then be correspondingly chosen – e.g., "=" if $Attr_i$ is categorical or ">" if $Attr_i$ is continuous.

In contrast, in the context of fuzzy rules Val_{ij} can take on either a nominal (discrete) or a *linguistic value*, depending on whether attribute $Attr_i$ is categorical or continuous. The comparison operator Op_i would then be correspondingly chosen – e.g., "=" if $Attr_i$ is categorical or "is" if $Attr_i$ is continuous.

condition 1		condition m
$Attr_1$ Op_1 Val_{1j} $Active_1$	$\cdot\ \cdot\ \cdot\ \cdot\ \cdot$	$Attr_m$ Op_m Val_{mj} $Active_m$

Figure 10.11: Fixed-length genotype encoding a variable-length rule antecedent

Now suppose that one allows Op_i to be only "=" (we do not allow Op_i to be other comparison operator such as "≠") in the case of categorical attributes. Then one can simplify the individual encoding by dropping the Op_i term from the genotype without loosing relevant information – as long as the EA knows which attributes are categorical and which were originally continuous (and are now being fuzzified). This holds because in this case both categorical and (originally-continuous) fuzzified attributes would always be associated with the same comparison operator, namely "=" and "is", respectively. Hence there is no need to include this redundant information into the genotype.

10.4.1.2 Determining the Consequent of a Fuzzy Rule

Once the rule antecedent is determined, it is necessary to choose the consequent of the rule. In a classification problem this is the class predicted by the rule.

There are several ways for determining the consequent of a fuzzy rule, similarly to the case of EAs for discovering crisp rules (see subsection 6.1.3). For instance, the consequent can be fixed and constant for each individual in the population [Bentley 2000], or chosen by a deterministic, elaborate procedure. Here we discuss only the latter approach in the context of fuzzy rules, since the former approach and its variations work in a similar way in both crisp and fuzzy cases.

In [Ishibuchi et al. 2000] a classification rule's consequent (a predicted class) is computed in three steps, as follows. Let $Antec$ be a given rule antecedent. First, the degree of matching between $Antec$ and each training data instance is computed. The basic idea of this computation was discussed in section 10.2, and will be discussed in somewhat more detail in subsection 10.4.2. Here we abstract away the details of this step and just illustrate the result of this step with a very simple example. Suppose the data set has just six data instances, four of them with class c_1 and two of them with class c_2. One possible result for this first step is shown in the table at Figure 10.12(a). This table has 6 rows, one for each data instance. The first column of this table indicates the instance class, while the second column indicates the degree of matching between the instance and $Antec$.

Second, for each class, compute the sum of the degrees of matching for all training data instances belonging to that class. Continuing our example, the result of this step is shown in the table at Figure 10.12(b). Third, the class assigned to the rule consequent is the class with the maximum value of the sum of degrees of matching computed in the previous step. Continuing our example, the result of this step is shown in Figure 10.12(c).

instance class	degree of matching
c_1	0.6
c_2	0.8
c_1	0.5
c_1	0.4
c_2	0.9
c_1	0.3

class	sum of degrees of matching
c_2	1.7
c_1	1.8

the class assigned to the
rule consequent is c_1,
since max(1.8,1.7) = 1.8

(a) first step   (b) second step   (c) third step

Figure 10.12: Example of the choice of a rule consequent for a fuzzy rule antecedent

The above procedure to choose the best rule consequent for a given rule antecedent seems intuitive, and it is essentially a generalization of the corresponding
procedure for a crisp rule. Indeed, in the case of crisp rules one chooses the majority class among all instances covered by the rule. This is just a simpler way of
saying that one computes the sum of "degrees of matching" (either 0 or 1 for a
crisp rule) for all training data instances belonging to a class and then chooses the
class with the maximum sum of degrees of matching.

However, in the case of fuzzy rules there is a caveat. There can be a reasonable number of data instances whose degree of matching can be relatively close to
0, say 0.1 or 0.2. Each of these data instances has a minor influence in the choice
of the class for a rule consequent, but collectively all these data instances can
have a significant influence in that choice. This may lead to some undesirable (or
at least not so intuitive) results.

For instance, let us revisit the choice of class c_1 in Figure 10.12. It could be
argued that a better choice of rule consequent would be c_2. Why? Note that the
two data instances with class c_2 in the table of Figure 10.12(a) are satisfying the
rule antecedent to a high degree, namely 0.8 and 0.9. In contrast, the four data
instances with class c_1 are satisfying the rule antecedent to relatively smaller degrees, namely 0.6, 0.5, 0.4, 0.3. Class c_1 was the "winner" simply because there
are more data instances of this class in the training set, and *not* because its data
instances have a better matching with the rule antecedent. In other words, the
above procedure has a bias favoring the choice of the majority class in the training set, and this bias tends to be undesirable when the class distribution is very
unbalanced.

Of course, a similar problem occurs when discovering crisp rules. In this case,
however, at least data instances that do not support the rule antecedent would
have a "degree of matching" of 0, so that the above problem would be somewhat
mitigated.

One possible solution for that problem, in the case of fuzzy rules, is *not* to
take into account data instances with a low degree of membership when computing the sum of the degrees of matching in the above second step of the procedure
for choosing the winner class. This idea can be realized by taking into account
only the data instances whose degree of matching is greater than or equal to a

predetermined threshold. In the example of Figure 10.12, if the threshold were 0.5 the sum of degrees of matching for class c_1 would be $0.6 + 0.5 = 1.1$, so that class c_2 would be the winner. This method was used in a non-evolutionary fuzzy-rule-discovery algorithm [Fertig et al. 1999].

10.4.2 Determining the Degree of Matching Between a Fuzzy Rule Antecedent and a Data Instance

The first step to compute this degree of matching is to compute the degree of membership of each attribute value of the data instance to the fuzzy sets representing attribute values that occur in the conditions of the rule antecedent. This computation is, of course, determined by the shape of the membership functions, as well as the number of membership functions (linguistic values) assigned to each attribute being fuzzified. Since in this section we assume that the membership functions of each attribute value are pre-specified, this is not an issue in the design of the EA itself. The issue of how to define membership functions will be discussed separately in subsection 10.4.4. Once all the necessary membership functions are defined, the degree of matching between an instance's attribute value and the corresponding rule condition is given by the value of the corresponding membership function, as discussed in section 10.2 – and illustrated in Figure 10.9.

Once the degree of membership of each rule condition has been computed, the second step is to compute the overall degree of membership of the rule antecedent as a whole. We assume that a rule antecedent consists of a conjunction of conditions, as usual. Hence, this step is determined by the kind of fuzzy AND operator chosen by the algorithm designer. There are several different fuzzy AND operators proposed in the literature, and there is no consensus on what is the "best" one. Furthermore, as usual in data mining and knowledge discovery, each of these operators has a distinct bias that favors a certain kind of rule over another, and the effectiveness of each of these bias is strongly dependent on the data being mined.

The product operator is a popular choice in the literature. It is used, e.g., in the algorithms proposed by [Ishibuchi et al. 2000; Ishibuchi and Nakashima 1999, 2000]. One characteristic of this kind of fuzzy AND operator is that the degree of membership of the whole rule antecedent tends to be quite sensitive to a very low or very high membership value of a single rule condition. For instance, suppose that a rule has three conditions in its antecedent, and that a given data instance satisfies each of these conditions to a degree of 0.7, 0.8 and 0.9, respectively. In this case the data instance satisfies the rule antecedent to a degree of $0.7 \times 0.8 \times 0.9 = 0.504$. Now suppose that there is some noise in the value of a single attribute in the data instance, so that the degree of membership to the third rule condition is mistakenly reduced from 0.9 to 0.1. In this case the degree of membership to the whole antecedent drops to 0.056.

This sensitivity to the degree of membership of a single attribute value is also present in crisp rules, of course. This kind of sensitivity is not necessarily a bad thing in general, but in some situations it may be undesirable – particularly when we know or suspect that the data is very noisy. In such cases we can instead use a fuzzy AND operator that is less sensitive to extreme values of degrees of membership in rule conditions.

An example of such operator is the one based on the median of the membership degrees of all rule conditions, as proposed by [Walter and Mohan 2000]. To compute the median of a list of values we first sort the list in ascending order. If the number of elements in the list is odd, the median is simply the middle-most value in the sorted list. If the number of elements in the list is even, the median is the arithmetic average of the two middle-most values. In the above example the median of the degrees of membership 0.7, 0.8, 0.9 is 0.8; whereas the median of 0.1, 0.7, 0.8 is 0.7. This illustrates that the median tends to be much less sensitive to extreme membership values in a single rule condition, in comparison with the product operator.

10.4.3 Using Fuzzy Rules to Classify a Data Instance

When using fuzzy rules to classify a new data instance, one often needs to combine the predictions of several fuzzy rules, or to select one of them, in order to decide which class will be assigned to a given data instance. This combination or selection can take into account several factors, such as:

1) The degree of matching associated with each rule. Intuitively, the higher the degree of matching between a rule and a data instance, the more influence that rule should have on the choice of the class assigned to that data instance.
2) The rule strength – computed as the confidence factor or another rule quality measure – of each rule. When this factor is taken into account, it is usually multiplied by degree of matching to give the final "firing strength" of the rule.

For instance, [Ishibuchi et al. 2000; Ishibuchi and Nakashima 2000] use a "single winner" approach, as follows. For each rule the algorithm computes the degree of matching between that rule and the instance to be classified, and then computes the product of this degree and the rule strength (confidence factor) of the rule. The rule with the maximum value of this product is chosen, and its predicted class is assigned to the data instance.

Rule strength is not considered by all EAs. For instance, [Walter and Mohan 2000] also use a single winner approach, but in their algorithm the choice of the winner rule is based only on the degree of matching between the data instance to be classified and each rule.

10.4.4 Specifying the Shape and the Number of Membership Functions

In the previous subsections we assumed the shape of membership functions, as well as the number of membership functions for each attribute, was previously specified, i.e., it was not evolved by the EA. In this subsection we discuss two approaches to specify the shape and number of membership functions, namely: (a) user-defined membership functions; and (b) clustering attribute values. Approaches (a) and (b) can be regarded as a form of manual and automatic preprocessing for the EA, respectively.

10.4.4.1 Manually Defined Membership Functions

The shape and number of membership functions for each attribute being fuzzified can be simply defined by the user. This approach is advocated, e.g., by [Ishibuchi and Nakashima 1999, 2000]. In this subsection we use the term "user" in a loose sense. Ideally, the person defining membership functions should be a domain expert, i.e., a person who is an expert on the target application domain. This person might or not be the ultimate user of the fuzzy rules discovered by the system. In any case, the basic idea of the approach discussed in this subsection is that membership functions are manually defined by a human being, rather than automatically defined by the system.

The motivation for this approach is twofold. First, it can be regarded as a form of incorporating some background knowledge of the user into the rule discovery system. The user can specify membership functions that are sensible, according to his knowledge of the application domain and the meaning of the data being mined. As a result, the membership functions will be consistent with the user's previous knowledge. It can be argued that, in general, consistency with the user's previous knowledge is one of the factors improving the comprehensibility of discovered knowledge [Pazzani 2000].

Second, this approach avoids the computationally-expensive process of trying to optimize the shape and number of membership functions. This problem becomes particularly serious as the number of attributes being fuzzified is increased. Note, however, that this reduction in the computer's processing time is achieved at the expense of increasing the time spent by the user in the knowledge discovery process.

Moreover, this approach also has other disadvantages. It reduces the degree of autonomy of the system, and the quality of the discovered rules depends on the quality of the membership functions specified by the user. In addition, different users may prefer membership functions that are quite different from each other. As a result, the system may have to be run several times, once for each user. This may significantly reduce the above-mentioned benefit of reducing computation time.

Methods for manually defining membership functions with the help of a domain expert or a set of experts are discussed in [Klir and Yuan 1995, chap. 10]. This reference discusses two general kinds of methods, namely:

(a) Direct methods, where experts give answers to questions directly related to the membership functions being defined. For example, a domain expert can directly assign to each element $x \in X$ a membership grade $\mu_A(x)$. The expert's answers are then used to define a membership function by using an appropriate curve-fitting method.
(b) Indirect methods, where experts answer simpler questions that are only indirectly related to membership functions. For example, instead of directly assigning a membership degree $\mu_A(x)$ to each element $x \in X$, an expert can simply compare elements x in a pairwise fashion according to their relative degree of membership in the fuzzy set A.

Indirect methods tend to require a more elaborate post-processing of the expert's answers, but, according to Klir and Yuan, they tend to be less sensitive to various biases of subjective judgement.

10.4.4.2 *Automatically Defined Membership Functions*

Instead of being manually-defined, as discussed in the previous subsection, membership functions can be automatically defined. One possible approach for automatically-defining membership functions consists of running a clustering algorithm. This approach has been used, e.g., in [Bentley 2000]. Other possible approach would be to use a neural network [Klir and Yuan 1995, pp. 295-300]. Here we focus on the use of clustering, whose basic ideas we have already introduced in section 2.4.

Recall that the goal of a clustering algorithm is to group together similar objects, by maximizing intra-cluster (within-cluster) similarity and minimizing inter-cluster (between-cluster) similarity.

For the purposes of this subsection, the "objects" to be clustered are simply the values of an attribute being fuzzified. In other words, in order to generate membership functions for an attribute a clustering algorithm is applied to a one-dimensional data subset, consisting of all values of that attribute observed in the training set. Each cluster of attribute values is then associated with a membership function.

Hence, the number of membership functions (and linguistic values) produced for each attribute can be either fixed a priori or dynamically adjusted to the data, depending on whether or not we use a clustering algorithm where the number of clusters is fixed in advance. A natural choice for the kind of clustering algorithm to be used is a fuzzy clustering algorithm, since in this case each object will be directly assigned to a cluster with a certain degree of membership.

10.5 Using EAs for Tuning Membership Functions

In this section we address the use of EAs for evolving only the membership functions associated with attributes being fuzzified, without evolving the fuzzy rules themselves.

This kind of EA is useful, for instance, when crisp prediction rules were already discovered by a conventional rule induction algorithm, and we just want to fuzzify the discovered crisp rules.

One example of an EA designed for this task is presented by [Crockett et al. 2000]. In this work a GA is used to fuzzify the decision nodes (the internal nodes) of a previously induced decision tree. In essence the GA optimizes the sizes of fuzzy regions around each decision node of the induced decision tree. For instance, consider the crisp decision tree shown in Figure 10.13.

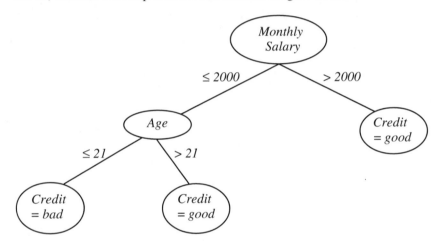

Figure 10.13: Example of induced decision tree

This decision tree has two decision (internal) nodes that are to be fuzzified. In this example the GA's goal is to define four membership functions, namely two opposing membership functions for the root node and two opposing membership functions for the node *Age*. The first two membership functions will fuzzify the meaning of the branches (*MonthlySalary* ≤ 2000) and (*MonthlySalary* > 2000), whereas the latter two membership functions will fuzzify the meaning of the branches (*Age* ≤ 21) and (*Age* > 21). Each membership function is determined by a gene of an individual, so that in this example four genes would be necessary to define the four membership functions.

Note that the number of membership functions (linguistic values) per attribute is fixed at 2, rather than optimized by the GA. The GA is used to optimize some aspect(s) related to the shape of each membership function. In the algorithm proposed by Crockett et al. each membership function is associated with a gene whose value is a parameter determining the lower bound and upper bound of the domain of the membership function, rather than the shape itself. The shape of each membership function is fixed for a given GA run, and the authors have experimented with several nonlinear functions, namely sigmoid, cubic, concave and convex curves.

More generally, one could specify a more complex individual encoding where each gene could determine not only the lower bound and upper bound of the domain of a membership function, but also the shape of that function.

In addition, other genes of the genotype of an individual can be used to define the fuzzy operators to be used in fuzzy inference. For instance, in Crockett et al.'s work an individual's genotype also includes two genes for defining parameterized fuzzy intersection and fuzzy union operators.

10.6 Using EAs for Both Generating Fuzzy Rules and Tuning Membership Functions

EAs can also be used to optimize both the contents of fuzzy rules (the combination of attribute values occurring in the rules) and the membership functions of the linguistic values of the attributes being fuzzified. In this case the genotype of an individual has at least two kinds of genes, so that it is useful to think of an individual as consisting of two strings, as illustrated in Figure 10.14.

linguistic value	. . . . .	linguistic value	memb. funct. parameters	. . . . .	memb. funct. parameters

genes defining contents of fuzzy rules   genes defining membership functions

Figure 10.14: Genotype for optimizing both rules and membership functions

Some EAs which are representative of this approach are the EAs proposed by [Pena-Reyes and Sipper 1999; Xiong and Litz 1999; Mota et al. 1999; Chen and Ho 2001]. The latter evolves fuzzy decision trees, rather than fuzzy rules. To simplify our discussion, we focus on the former two EAs. (For a review of EAs optimizing both rules and membership functions see also [Romao et al. 2000].)

Both GAs follow the Pittsburgh approach, where each individual encodes a set of fuzzy rules. At a high level of abstraction, in both algorithms the genotype of an individual can be thought of as consisting of two parts, as in Figure 10.14.

Let us first review the basic ideas of the GA proposed by [Pena-Reyes and Sipper 1999], and later on we will point out some differences between this algorithm and the one proposed by [Xiong and Litz 1999].

The former algorithm was introduced in the context of a particular classification problem, where the goal was to discover fuzzy rules for breast cancer diagnosis, but its basic ideas are of course applicable to other classification problems.

In this GA the first part of the genotype defines the attribute values occurring in the antecedents of the rules contained in an individual. The consequent is the same for all rules in all individuals of the population, i.e., it is fixed for the entire GA run. Assuming that all predictor attributes are being fuzzified, a rule antecedent is a conjunction of conditions of the form:

$$\text{IF } (Attr_1 \text{ is } Val_{1j}) \text{ AND ... AND } (Attr_m \text{ is } Val_{mj}),$$

where m is the number of predictor attributes, $Attr_i$ denotes the i-th attribute, $i=1,...m$, and Val_{ij} denotes the j-th value of the domain of $Attr_i$. Each attribute domain includes the "don't care" symbol, indicating that the corresponding condition does not occur in the rule antecedent. All the other values of an attribute's domain are linguistic values.

For each rule in the genotype, each Val_{ij} can be encoded by a few bits or by a higher-level representation, where an integer number directly specifies the j-th linguistic value. (In [Pena-Reyes and Sipper 1999] each V_{ij} was encoded by two bits.)

The second part of the genotype defines, for each attribute being fuzzified, two parameters that specify the exact shape of the two trapezoidal membership functions shown in Figure 10.15. The two parameters in question are called P and d in the figure. Each of these two parameters has its value encoded by 3 bits, so that the total length of this part of the genotype is: 3 bits × 2 × number_of_fuzzified_attributes. Of course, other kinds of membership functions with a different number of parameters, and/or a different number of bits per parameter, could be used, resulting in a different genotype length.

The GA proposed by [Xiong and Litz 1999] – hereafter called X&L for short – differs from the previous GA – hereafter called PR&S for short – in some aspects; though, as mentioned above, both algorithms are similar at a high level of abstraction. Some differences are summarized in the following.

PR&S uses a binary encoding for both parts of the genotype – defining the contents of the fuzzy rules and the parameters of membership functions. In contrast, X&L uses a binary encoding for the fuzzy rules and an integer coding for the membership functions. The latter is used instead of binary encoding in order to reduce the length of an individual's genotype.

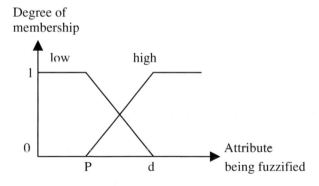

Figure 10.15: Membership function used by [Pena-Reyes and Sipper 1999]

In addition, X&L uses a "three-point" crossover, where the second crossover point is always fixed, falling on the boundary between the two substrings. The first and third crossover point fall inside the first and second substring, respectively. This is equivalent to perform a one-point crossover with each pair of corresponding (homologous) substrings. In other words, the genetic material in each substring is crossed-over separately, independent of the genetic material in the other substring. Apparently this scheme is not used in PR&S.

Another important difference between the two algorithms concerns the way in which the rule consequent is defined. In PR&S the rule consequent is fixed during a GA run, i.e., all rules of all individuals have the same consequent. In contrast, in X&L the rule consequent is deterministically chosen from the candidate consequents (classes) in such a way that the truth value of the rule as a whole is maximized.

The two above-described algorithms follow the approach of evolving a single population, where each individual's genotype has two parts, one specifying contents of fuzzy rules and the other one specifying membership function parameters. There is another approach for using an EA for both generating fuzzy rules and tuning membership functions. This alternative approach consists of using co-evolution, where two populations co-evolve to try to find a synergistic combination of fuzzy rules (evolved in one population) and fuzzy membership functions (evolved in another population).

A co-evolutionary system that is representative of this approach is the system proposed by [Mendes et al. 2001]. In essence, this system consists of two co-evolving EAs, each of them associated with a distinct population. The first one is a Genetic Programming (GP) algorithm where each individual of the population represents a fuzzy rule set. A GP individual specifies only which attribute values occur in its rules. The definitions of the membership functions necessary to interpret the fuzzy rule conditions represented by an individual are provided by the second population.

The second algorithm is a simple EA, which works with a "population" of a single individual. The individual in the current population undergoes mutation, but not crossover. This single individual specifies definitions of all the membership functions for all attributes being fuzzified. These definitions are used by the first population of GP individuals.

For instance, one of the rules of an individual could contain the fuzzy condition ($Age = old$), where old is a linguistic value. The definition of the membership function of old for the attribute Age would be provided by the second population.

Hence, the first population provides fuzzy rule sets for the second population, and those fuzzy rule sets are used to compute the fitness of the membership function definitions being evolved by the second population. Conversely, the second population provides membership function definitions for the first population, and those membership function definitions are used to compute the fitness of the fuzzy rule sets being evolved by the first population. As a result, the system simultaneously evolves both fuzzy rule sets and membership function definitions that are well adapted to each other.

The main advantage of this co-evolutionary approach seems to be a greater robustness in the evaluation of fitness functions. In particular, the fitness of a given set of membership function definitions is evaluated across several fuzzy rule sets, represented by several different GP individuals, rather than on a single fuzzy rule set. Of course, this advantage is obtained at the expense of increasing processing time. In order to cope with this problem, the authors use the approach of evaluating a given set of membership functions (the individual of the second population) only across the few best fuzzy rule sets (GP individuals of the first population) of the current generation.

10.7 Fuzzy Fitness Evaluation

We have discussed in section 6.5 how the fitness of an EA for rule discovery can be based on a confusion matrix, which summarizes the performance of a classification rule with respect to predictive accuracy. For convenience of the reader, we reproduce in Figure 10.16 the basic structure of a confusion matrix, which is the same as Figure 6.9.

actual class

		c	not c
predicted	c	TP	FP
class	not c	FN	TN

Figure 10.16: Confusion matrix for a classification rule

Recall that in this figure c denotes the class predicted by a rule, which is also called the positive class. The class predicted for a given data instance is c if that instance satisfies the rule antecedent, denoted by A. Otherwise the instance is assumed to belong to the negative class.

The labels in each quadrant of the matrix have the following meaning:

TP (True Positives) = Number of instances satisfying A and having class c
FP (False Positives) = Number of instances satisfying A but not having class c
FN (False Negatives) = Number of instances not satisfying A but having class c
TN (True Negatives) = Number of instances not satisfying A nor having class c

In section 6.5 we have implicitly assumed that TP, FP, FN and TN are crisp variables, which take on an integer value. However, this assumption is not valid in the context of fuzzy rules, as discussed in this chapter. In this context a data instance can satisfy a rule antecedent to a certain (real-valued) degree in the range [0..1], which corresponds to the degree of membership of that instance in that rule antecedent.

Therefore, it seems natural to fuzzify the confusion matrix shown in Figure 10.16, computing fuzzy values for TP, FP, FN and TN [Mendes et al. 2001]. In this case these variables have the following meaning:

TP (True Positives) = Summation of the degrees of membership in A for all in-
          stances that have class c;

FP (False Positives) = Summation of the degrees of membership in A for all in-
          stances that do not have class c;

FN (False Negatives) = Summation of the degrees of membership in (*not A*) for
          all instances that have class c;

TN (True Negatives) = Summation of the degrees of membership in (*not A*) for
          all instances that do not have class c;

To illustrate the computation of a fuzzy confusion matrix, suppose that a given data instance satisfies the antecedent A of the rule being evaluated to a degree of 0.8. Suppose also that this data instance has the class c predicted by the rule. Since the instance's class and the rule's predicted class match, ideally one would like the instance to satisfy the rule to a degree of 1. Since the instance is satisfying the rule to a degree of 0.8, one can say that the rule is 80% correct and 20% wrong. Therefore, this instance contributes a value of 0.8 for the summation of membership degrees constituting the value of TP and contributes a value of 0.2 for the summation of membership degrees constituting the value of FN. (We are assuming that *not A* is defined by the standard complement of the fuzzy set A, i.e., $\mu_{NOT-A}(x) = 1 - \mu_A(x)$, as discussed in subsection 10.1.2.)

As another example, suppose that a given data instance satisfies the antecedent A of the rule being evaluated to a degree of 0.3, and that this instance does not have the class c predicted by the rule. Since the instance's class and the rule's predicted class do not match, ideally one would like the instance to satisfy the rule to a degree of 0. Since the instance is satisfying the rule to a degree of 0.3, one can say that the rule is 70% correct and 30% wrong. Therefore, this instance contributes a value of 0.7 for the summation of membership degrees constituting the value of TN and contributes a value of 0.3 for the summation of membership degrees constituting the value of FP.

Once the fuzzy computation matrix associated with a rule (or rule set) has been computed as discussed above, the EA can use any of the fitness functions discussed in section 6.5 to evaluate the quality of a rule or a rule set.

In any case, it should be noted that fuzzifying a confusion matrix is not the only possibility to evaluate fuzzy rules. Alternatively, one could work with a crisp confusion matrix, by defuzzifying a rule when its fitness is computed. This is the approach followed by [Dasgupta and Gonzales 2001]. In this work, first the degree of matching between the rule antecedent and a data instance is computed. Now there are two possibilities: (a) If this degree is greater than or equal to 0.5 then the instance satisfies the rule antecedent in a crisp sense – the instance will contribute a value of 1 for either the TP or FP cell of the confusion matrix, depending on the instance's class. (b) If the degree of matching between the instance and the rule antecedent is less than 0.5 then the instance does not satisfy the rule antecedent in a crisp sense – the instance will contribute a value of 1 for either the FN or TN cell of the confusion matrix, depending on the instance's class.

References

[Bentley 2000] P.J. Bentley. "Evolutionary, my dear Watson" – investigating committee-based evolution of fuzzy rules for the detection of suspicious insurance claims. *Proceedings of the Genetic and Evolutionary Computation Conference (GECCO '2000)*, 702–709. Morgan Kaufmann, 2000.

[Chen and Ho 2001] H.-M. Chen and S.-Y. Ho. Designing an optimal evolutionary fuzzy decision tree for data mining. *Proceedings of the Genetic and Evolutionary Computation Conference (GECCO '2001)*, 943–950. Morgan Kaufmann, 2001.

[Cox 1995] E. D. Cox. *Fuzzy Logic for Business and Industry*. Charles River Media, 1995.

[Crockett et al. 2000] K.A. Crockett, Z. Bandar and A. Al-Attar. Soft decision trees: a new approach using non-linear fuzzification. *Proceedings of the 9th IEEE International Conference Fuzzy Systems (FUZZ IEEE '2000)*, 209–215. San Antonio, TX, USA, May 2000.

[Dasgupta and Gonzales 2001] D. Dasgupta and F.A. Gonzales. Evolving complex fuzzy classifier rules using a linear tree genetic representation. *Proceedings of the Genetic and Evolutionary Computation Conference (GECCO '2001)*, 299–305. Morgan Kaufmann, 2001.

[Espinosa and Vandewalle 2000] J. Espinosa and J. Vandewalle. Constructing fuzzy models with linguistic integrity from numerical data – AFRELI algorithm. *IEEE Transactions on Fuzzy Systems 8(5)*, 591–600, 2000.

[Fertig et al. 1999] C.S. Fertig, A.A. Freitas, L.V.R. Arruda and C. Kaestner. A Fuzzy Beam-Search Rule Induction Algorithm. *Principles of Data Mining and Knowledge Discovery (Proceedings of the 3rd European Conference – PKDD '99). Lecture Notes in Artificial Intelligence 1704*, 341–347. Springer, 1999.

[Hirota and Pedrycz 1999] K. Hirota and W. Pedrycz. Fuzzy computing for data mining. *Proceedings of the IEEE, 87(9)*, 1575–1600, 1999.

[Ishibuchi and Nakashima 1999] H. Ishibuchi and T. Nakashima. Designing compact fuzzy rule-based systems with default hierarchies for linguistic approximation. *Proceedings of the Congress on Evolutionary Computation (CEC '99)*, 2341–2348. Washington, D.C., 1999.

[Ishibuchi and Nakashima 2000] H. Ishibuchi and T. Nakashima. Linguistic rule extraction by genetics-based machine learning. *Proceedings of the Genetic and Evolutionary Computation Conference (GECCO '2000)*, 195–202. Morgan Kaufmann, 2000.

[Ishibuchi et al. 2000] H. Ishibuchi, T. Nakashima and T. Kuroda. A hybrid fuzzy GBML algorithm for designing compact fuzzy rule-based classification systems. *Proceedings of the 9th IEEE International Conference Fuzzy Systems (FUZZ IEEE 2000)*, 706–711. San Antonio, TX, USA, 2000.

[Janikow 1995] C.Z. Janikow. A genetic algorithm for optimizing fuzzy decision trees. *Proceedings of the 6th International Conference on Genetic Algorithms (ICGA '95)*, 421–428. 1995.

[Klir and Yuan 1995] G.J. Klir and B. Yuan. *Fuzzy Sets and Fuzzy Logic: Theory and Applications*. Prentice-Hall, 1995.

[Kosko 1994] B. Kosko. *Fuzzy Thinking: the New Science of Fuzzy Logic.* Flamingo, 1994.

[Mendes et al. 2001] R.F. Mendes, F.B. Voznika, A.A. Freitas and J.C. Nievola. Discovering fuzzy classification rules with genetic programming and co-evolution. *Principles of Data Mining and Knowledge Discovery (Proceedings of the 5th European Conference, PKDD '2001) – Lecture Notes in Artificial Intelligence 2168*, 314–325, Springer, 2001.

[Mota et al. 1999] C. Mota, H. Ferreira and A. Rosa. Independent and simultaneous evolution of fuzzy sleep classifiers by genetic algorithms. *Proceedings of the Genetic and Evolutionary Computation Conference (GECCO '99)*, 1622–1629. Morgan Kaufmann, 1999.

[Pazzani 2000] M.J. Pazzani. Knowledge discovery from data? *IEEE Intelligent Systems, 15(2)*, 10–13, 2000.

[Pena-Reyes and Sipper 1999] C.A. Pena-Reyes and M. Sipper. Designing breast cancer diagnostic systems via a hybrid fuzzy-genetic methodology. *Proceedings of the 8th IEEE International Conference Fuzzy Systems (FUZZ IEEE'99)*. IEEE, 1999.

[Romao et al. 2000] W. Romao, A.A. Freitas and R.C.S. Pacheco. Uma revisao de abordagens genetico-difusas para descoberta de conhecimento em banco de dados. (In Portuguese) *Acta Scientiarum 22(5)*, 1347–1359. Universidade Estadual de Maringa, Brazil, 2000.

[Setnes and Roubos 2000] M. Setnes and H. Roubos. GA-Fuzzy modeling and classification: complexity and performance. *IEEE Transactions on Fuzzy Systems 8(5)*, 509–522, 2000.

[Walter and Mohan 2000] D. Walter and C.K. Mohan. ClaDia: a fuzzy classifier system for disease diagnosis. *Proceedings of the Congress on Evolutionary Computation (CEC '2000)*. La Jolla, CA, USA, 2000.

[Xiong and Litz 1999] N. Xiong and L. Litz. Generating linguistic fuzzy rules for pattern classification with genetic algorithms. *Principles of Data Mining and Knowledge Discovery (Proceedings of the 3rd European Conference, PKDD '99). Lecture Notes in Artificial Intelligence 1704*, 574–579. Springer, 1999.

[Yen 1999] J. Yen. Fuzzy logic – a modern perspective. *IEEE Transactions on Knowledge and Data Engineering 11(1)*, 153–165, 1999.

11 Scaling up Evolutionary Algorithms for Large Data Sets

> "In a world where serial algorithms are usually made parallel
> through countless tricks and contortions, it is no small irony
> that genetic algorithms (highly parallel algorithms) are
> made serial through equally unnatural tricks and turns."
> [Goldberg 1989, p. 208]

One well-known disadvantage of evolutionary algorithms (EAs) for rule discovery is that in general they are slow, by comparison with rule discovery algorithms based on the rule induction paradigm. After all, rule induction algorithms usually perform a kind of local search in the rule space, whereas EAs are population-based algorithms that perform a more global search of the rule space.

Therefore, when the data to be mined is very large it is necessary to speed up EAs. This chapter discusses two basic approaches for speeding up EAs, scaling them up for large data sets. The first approach, discussed in section 11.1, consists of using only a subset of the available data to evaluate the fitness of an individual. The second approach, discussed in section 11.3, consists of parallelizing EAs and then running them on a parallel computer. For readers who are not familiar with parallel processing techniques, section 11.2 presents an overview of parallel processing that will be useful for a better understanding of section 11.3.

11.1 Using Data Subsets in Fitness Evaluation

In data mining applications (as well as in most applications) the most computationally expensive procedure performed by an EA is usually fitness evaluation. The problem is that in a conventional EA for data mining the fitness of each individual must be computed by scanning all data instances of the data being mined – the entire training set, in the case of classification or another task involving prediction, or the entire data set, in the case of clustering or another task without separation between training and test sets. As a result, when the data being mined is very large the EA will be very slow.

In this section we discuss one approach for speeding up EAs based on using only subsets of data instances of the data being mined for fitness evaluation. We focus on the classification task. The basic idea is that, when the data set being mined is very large, evaluating fitness of individuals (candidate rules or rule sets) on a subset of training instances, rather on all training instances, will in general significantly speed up the EA. Hopefully, this significant speed up will be obtained without unduly reducing the predictive accuracy or another quality measure of the discovered rules.

We will first discuss the approach where the subset of training instances used for fitness evaluation is randomly selected. We will refer to this approach as random training-subset selection. Then we will discuss the approach where the subset of training instances used for fitness evaluation is dynamically adapted during evolution. We will refer to this approach as adaptive training-subset selection.

11.1.1 Random Training-Subset Selection

Random training-subset selection is actually a form of resampling. Each random sample can be extracted from the training set either with or without replacement. In any case, in the context of EAs for data mining one can distinguish among at least three possibilities for defining the frequency of resampling [Bhattacharyya 1998; Cavaretta and Chellapilla 1999], as follows:

(a) *Individual-wise*: In this approach a new sample of data instances is extracted from the training set for each individual of the population. As a result, different individuals will probably be evaluated on different data samples. This has the disadvantage of casting some doubts on the fairness of the selection procedure of the EA.

(b) *Run-wise*: In this approach a single fixed sample of data instances is extracted from the training set and used to evaluate the fitness of all individuals throughout the EA run. This has the disadvantage that a small, fixed sample would probably significantly reduce the robustness and predictive accuracy of the rules discovered by the EA. In this approach there is a particularly challenging trade-off between speeding up the EA (reducing processing time) and maintaining the predictive accuracy as high as possible. The "optimal" value of the sample size to cope with this trade-off tends to be strongly problem-dependent.

(c) *Generation-wise*: In this approach a fixed sample of data instances is extracted from the training set at each generation, and all individuals of that generation have their fitness evaluated on that data sample. This approach avoids the disadvantages of the two previous approach, and so it seems more effective. In particular, note that an individual will survive for several generations only if its corresponding rule or rule set has a good predictive accuracy across different data samples. Hence, this approach favors the discovery of rules that are more robust across different data samples, i.e., rules that seem to have a better generalization ability.

The idea of selecting a subset of training instances to speed up fitness evaluation is not restricted to EAs that discover prediction rules. For instance, [Sharpe and Glover 1999] use random instance-subset selection in a GA for attribute selection. This GA uses the standard individual encoding for attribute selection, as discussed in subsection 9.1.1.1. That is, each individual contain m genes, where m is the number of original attributes, and each gene can take on the value "1" or "0", indicating whether or not its corresponding attribute is selected, respectively. The GA follows the wrapper approach, where the fitness of an individual is computed by evaluating the performance of a classification algorithm on a training-test set. This is a part of the training set separated from the training-

training set, which is used to train the classification algorithm with the attributes selected by the individual. (See subsection 4.1.2 for a review of these concepts.)

In this work a random instance-subset selection method was used to select a subset of instances from the training-test set (called the "test" set in the terminology of the authors). When the evolution of the GA is completed, the best individual (attribute subset) returned by the GA can be determined as the highest-fitness individual generated by the GA, as usual. However, if the size of the subset of instances used in fitness evaluation was small then fitness values may not be a very reliable indicator of an individual's quality. To mitigate this problem the authors propose two alternative methods of determining the individual returned by the GA.

One method consists of using the entire training-test set (rather than just a subset of it) to evaluate the fitness of all individuals of the last generation. The highest-fitness individual, as measured on the entire training-test set, is then selected as the individual returned by the GA. This is a sensible approach. It allocates more time for computing the fitness of individuals of the last generation, since these individuals tend to be better solutions than individuals from previous generations. Note that this idea is not restricted to the attribute selection task. It is generic enough to be used in any EA for data mining where fitness evaluation is based on a small data sample in order to speed up the EA.

Another method proposed by the authors consists of forming a composite individual by combining information from all individuals of the last generation. The method examines each gene (bit) in turn for all those individuals. The value of each gene is set to "1" if the majority of individuals of the last generation have the value "1" for that gene, and set to "0" otherwise. This method is less computationally expensive than the previous one, since it does not re-compute the fitness of individuals of the last generation on the entire training-test set. Both methods gave similar results, which were better than the results obtained by using the conventional approach of making the GA simply return the highest-fitness individual (measured on a sample of the training-test set) generated by the GA.

11.1.2 Adaptive Training-Subset Selection

Instead of selecting a training subset in a random way, as discussed in the previous subsection, one can use an adaptive approach to dynamically select a training subset for fitness evaluation. The motivation for this approach is not only to speed up the EA, but also to increase the effectiveness of the sampling procedure by dynamically selecting "difficult" training instances, i.e., instances that are frequently misclassified. Quoting [Gathercole and Ross 1997a, p. 119]:

"Simply evaluating each individual in each generation on the entire training set ignores the current abilities of the population which change with each generation, wasting a good resource ... and ignores information gained about the difficulty of cases in the training set."

[Gathercole and Ross 1994] proposed three different training-subset selection methods. One of them was a simple generation-wise, random subset selection method, similar to the one discussed in the previous subsection. At each generation a sample of training data instances is selected for fitness evaluation, for all individuals of that generation. Data instances are selected entirely at random, so that each data instance has the same probability of being selected. In contrast, the two other methods used in this work are adaptive methods, as follows.

The Historical Subset Selection method consists of using the results of previous EA runs to select difficult (frequently-misclassified) instances. More precisely, the EA is run a few times (about five times), and the instances that are misclassified by the best individual of each generation of each run are recorded. These instances are then used to make up the training subset in further EA runs. Apparently this training subset remains fixed for an entire run after its initial selection, characterizing a run-wise training-subset selection approach (in the sense discussed in the previous subsection).

The Dynamic Subset Selection method is based on the idea of dynamically selecting instances that are difficult and/or have not been selected for several generations. This is a generation-wise training-subset selection method. At each generation a subset of training data instances is chosen by a biased selection procedure. More precisely, the probability of an instance being selected increases with measures of its "difficulty" and its "age". In essence, the difficulty of an instance is measured by the number of times that the instance is misclassified by the EA individuals, whereas the age of an instance is measured by the number of generations since the instance was last selected. The measures of difficulty and age of an instance are then combined into a single formula defining the instance's probability of being selected. This is a weighted formula, and the relative weights of difficulty and age are parameters of the algorithm.

Another method for adaptive training-subset selection, called Limited Error Fitness, was proposed by [Gathercole and Ross 1997a]. In this method at each generation the data instances of the training set are re-ordered, so that the training "set" actually becomes an ordered list of data instances. In general the re-ordering procedure aims at moving the easiest data instances to the end of the ordered training list, so that the most difficult instances will be moved to the beginning of the ordered training list.

The fitness of an individual is determined by how many of the ordered training instances it classifies correctly before it makes a certain number of misclassifications, called the error limit. After exceeding the error limit, any training instance that is not covered by the individual is counted as a misclassified instance, in order to speed up fitness computation. The fitness of an individual is the total number of instances misclassified by that individual.

Both a re-ordering of training instances and the value of the error limit are dynamically determined at each generation based on the performance of the best individual from the previous generation.

Limited Error Fitness can be considered as a hybrid generationwise/individual-wise training-subset selection method, in the following sense. The re-ordering of training instances and determination of error limit are made on a kind of generation-wise basis. However, the actual number of instances to be classified by each individual can vary a lot between different individuals, depending on their quality. More precisely, individuals representing better rules have their fitness evaluated by actually classifying a larger number of training instances, whereas individuals representing worse rules have their fitness evaluated by actually classifying a smaller number of training instances. In contrast, in a pure generation-wise method the same instance subset would be used to actually evaluate all individuals of the current generation, independent of each individual's quality. An experimental comparison between Dynamic Subset Selection and Limited Error Fitness is reported in [Gathercole and Ross 1997b]. Limited Error Fitness was successfully used in [Papagelis and Kalles 2001], but the authors do not give details about the use of this technique.

[Sharpe and Glover 1999] proposed an adaptive data sampling method based on a combination of statistical significance testing and the principle of Occam's razor (the latter is discussed in subsection 2.5.2). The method works with tournament selection. Two individuals are picked at random from the population to compete in a tournament. Instead of evaluating the fitness of those individuals on all available data instances, their fitness is evaluated on a subset of data instances. Data instances are added to this subset until either a maximum number of data instances is reached or a clear tournament winner emerges. Whether or not the latter condition is satisfied is determined by a statistical significance test. Hence, the size of the subset of instances used in fitness computation is dynamically adapted to the relative quality of the two individuals competing in the tournament. (A conceptually similar idea was also proposed by [Teller and Andre 1997].)

If no clear tournament winner emerges when a maximum number of data instances is reached, then the simplest solution (represented by an individual) is selected, based on Occam's razor. In Sharpe and Glover's work the EA is actually a GA where each individual represents a subset of selected attributes (as discussed at the end of the previous subsection), so that the simplest solution is represented by the individual with the smallest number of selected attributes.

11.2 An Overview of Parallel Processing

In this section we present an overview of the main concepts of the area of parallel processing that are necessary for a better understanding of the next section, where we discuss how to parallelize EAs for data mining. The discussion presented in this section is based on chapters 6 and 7 of [Freitas and Lavington 1998].

11.2.1 Basic Concepts of Parallel Processing

In essence, parallel processing involves the simultaneous execution of tasks by several processors. At a given instant the processors can be executing the same task or different tasks. In the former case, although each processor is executing the same task (or instruction), different processors are usually processing different data. Hence, this kind of parallel processing is often called Single Instruction stream, Multiple Data stream (SIMD). In contrast, in the latter case in general different processors are executing different tasks on different data. Hence, this kind of parallel processing is often called Multiple Instruction stream, Multiple Data stream (MIMD).

Within the MIMD category there are essentially two major kinds of systems with respect to memory distribution, namely shared-memory and distributed-memory systems. The basic idea of each of these two kinds of system is shown in Figure 11.1, where each processor is denoted by p and each local memory (in the case of distributed memory) is denoted by M. In a shared-memory system each processor has direct access to a global address space, as shown in Figure 11.1(a). Hence, all inter-processor communication occurs via access to the global memory. In contrast, in a distributed-memory system each processor has direct access only to its local memory, as shown in Figure 11.1(b). Hence, all inter-processor communication occurs via message passing through an interconnection network.

It should be noted that a global, shared memory is a much more powerful communication mechanism than a distributed memory. In particular, when using a shared-memory system the programmer does not need to worry about which processor's local memory contain which data, since all data is directly available to all processors. Furthermore, the programmer does not need to worry about how to minimize the amount of inter-processor communication on the underlying interconnection-network topology. In addition, in a shared memory architecture the crucial problem of load balancing (discussed in the next subsection) is greatly simplified.

However, the advantages of shared-memory systems are not obtained for free. Shared-memory systems tend to be more expensive and less scalable for a large number of processors, due, e.g., to an increasing difficulty in coping with memory contention as the number of processors is increased.

Actually, shared-memory and distributed-memory architectures can be considered as two extreme kinds of memory organization. In practice there are also hybrid organizations, using some kind of logically-centralized global address space but with a physically-distributed memory. In this kind of system, although each processor has access to the entire global address space, the time that a processor takes to access data in a remote processor's local memory is typically longer than the time that the processor takes to access data in its own local memory.

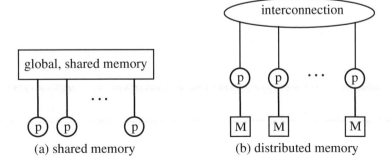

(a) shared memory   (b) distributed memory

Figure 11.1: Basic difference between shared memory and distributed memory

Two other basic concepts of parallel processing are the related concepts of granularity and level of parallelism [Almasi and Gottlieb 1994]. In essence, the granularity is the average size of the tasks assigned to the processors, whereas the level of parallelism is the level of abstraction at which parallelism is exploited. Fine-grain tasks are associated with a low level of parallelism (e.g., at the instruction level), whereas coarse-grain tasks are associated with a higher level of parallelism (e.g., at the procedure level or program level).

11.2.2 Load Balancing

The workload of a parallel program consists of several tasks, which are distributed across the available processors of a parallel computer. Ideally this workload should be distributed across the processors in such a way that the total processing time of the program is minimized. This is essentially the problem of load balancing.

In this section we assume that all processors are homogeneous, which is a common case in practice. In this case a good solution for the load balancing problem consists of distributing tasks among the processors in such a way that each processor receives approximately the same amount of workload.

(Of course, if the processors were heterogeneous – they had different processing capacities, such as different clock rates – intuitively one should assign more workload to faster processors and less workload to slower processors, in order to achieve a better load balancing.)

Assigning approximately the same amount of workload to each processor of a set of homogenous processors is important because the time taken by a set of processors working in parallel will be the time taken by the processor with the largest workload. For instance, suppose a workload is divided into four tasks, t_1, t_2, t_3 and t_4, which are to be distributed across two homogeneous processors p_1 and p_2. Suppose these tasks take 1, 2, 3 and 4 minutes, respectively. A bad solution for the load balancing problem would be to assign tasks t_1 and t_2 to processor p_1 and tasks t_3 and t_4 to processor p_2. In this case the total processing time would be 7 minutes, which would be the time taken by p_2. After completing its two tasks in 3 minutes processor p_1 would be idle for 4 minutes. Clearly a better solution is to assign tasks t_1 and t_4 to processor p_1 and tasks t_2 and t_3 to processor p_2. In this case both processors complete their tasks in 5 minutes and no processor gets idle.

As mentioned above, the problem of load balancing is greatly simplified in a shared-memory system, where each processor has direct access to a global memory. In this case the system can easily perform load balancing by assigning the next task (in a queue of tasks) to the first processor that becomes available.

In contrast, load balancing is a difficult problem in distributed-memory systems. In this kind of system inter-processor communication can take a considerable time. Hence, a load balancing strategy should try to minimize inter-processor communication, which is a goal in conflict with the goal of improving load balancing. (One could minimize inter-processor communication by distributing all workload to a few processors out of all available processors, but this would lead to a poor load balancing solution.) In addition, the performance of a given load balancing strategy tends to be dependent on the topology of the interconnection network.

In the context of data mining and other data-intensive applications, load balancing in distributed-memory systems is further complicated by the fact that not only the tasks or instructions but also the data being processed is distributed across the available processors. On one hand one would like to evenly distribute data across the local memories of all processors, to avoid data skew. On the other hand, minimizing data skew does not necessarily minimize processing skew. It is often the case that different tasks have to process different data subsets. To see the difficulty of load balancing in data-intensive applications, consider a situation where the different data subsets to be processed by different tasks have very different sizes. In this case, if one evenly distributes the entire data set across the local memories of all available processors, one would achieve a good "data balancing" solution (minimizing data skew), but probably a bad "processing balancing" solution (increasing processing skew).

11.2.3 Data Parallelism vs Control Parallelism

In this subsection we discuss two broad kinds of parallelism, namely control parallelism and data parallelism [Freitas and Lavington 1998, chap. 7; Freitas 1998]. Understanding the crucial differences between these two kinds of parallelism is particularly important to better understand how to parallelize an EA for data mining, which will be discussed in section 11.3.

In essence, data parallelism involves the execution of the same procedure on multiple large data subsets at the same time [Hillis and Steele 1986; Lewis 1991]. This approach is illustrated in Figure 11.2. As shown in this figure, the data set being processed is divided into p data subsets, where p is the number of available processors. Each processor applies the same procedure to its local data subset, computing a partial result. All p processors work in parallel, at the same time. Then the partial, local results of all p processors are combined into a single global result.

In contrast, control parallelism involves the concurrent execution of multiple different procedures. This approach is illustrated in Figure 11.3. At a given time instant each of the p processors is executing a different procedure on its corresponding data set.

Note that when exploiting data parallelism the flow of control of the algorithm is essentially sequential, in the sense that the algorithm executes one procedure at a time. Only data-manipulation procedures are parallelized. This leads to some advantages of data parallelism. In particular, the flow of control of sequential algorithms can be re-used, avoiding the need for explicit parallelization of that flow of control.

In contrast, when exploiting control parallelism the flow of control of the algorithm must be carefully parallelized. Moreover, the effectiveness of a given parallelization strategy usually depends on several aspects of the underlying architecture of the parallel system, e.g., on the topology of the interconnection network linking the processors of a distributed-memory system.

In other words, exploiting data parallelism tends to be simpler and more independent of the architecture of the parallel system being used than exploiting control parallelism.

In addition, in the context of data mining and other data intensive applications, data parallelism has the advantage of being more scalable with respect of the size of the data being mined – see subsection 11.3.2.

Finally, note that hybrid data-parallelism and control-parallelism approaches can be used in an attempt to get the best of both worlds. The key for developing a hybrid approach is to realize that data parallelism and control parallelism actually address different kinds of "large" problems, as follows.

Data parallelism addresses the problem of manipulating very large data sets, since it involves the parallelization of data-manipulation operations. In the context of an EA for rule discovery, this means that data parallelism is particularly useful for parallelizing the computation of a fitness function when mining a very large data set.

Control parallelism addresses the problem of performing a search in a very large search space. In the context of an EA for rule discovery, this means that control parallelism is particularly useful for processing several different individuals (candidate rules or rule sets) in parallel, at the same time.

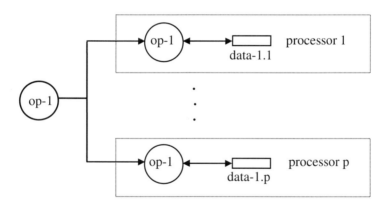

Figure 11.2: data parallelism

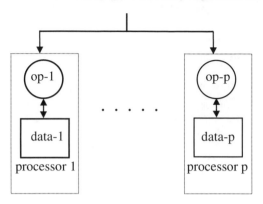

Figure 11.3: control parallelism

11.2.4 Speed up and Efficiency Measures

Two simple and commonly used measures of performance of a parallel, multiple-processor system are the Speed Up (Sp) and the Efficiency (Ef) of the system by comparison with a sequential, single-processor system. Before we define these measures, it is important to notice that in general they will be fair measures of performance of a parallel system only when two conditions hold, namely:

(a) All the processors and other hardware resources (e.g., main memory) of the parallel system are homogenous and are exactly of the same kind as their counterparts in the sequential system being compared with the parallel system.

(b) Let p be the number of processors of the parallel system. Then all hardware resources of the parallel system (e.g., amount of main memory) should be p times as much as in the sequential system being compared with the parallel system.

In essence, the Sp of a parallel system with p processors over a sequential system with a single processor can be defined as $Sp = T_1 / T_p$, where T_1 is the time taken to run a program on a single processor system and T_p is the time taken to run a parallel version of that program on the system with p processors. For instance, if a program takes 10 minutes to run on a sequential system and a parallel version of that program takes 2 minutes to run on a system with 10 processors, then the Sp would be 5 (10/2).

Ideally one would like a parallel system to achieve a linear speed up, so that a p-fold increase in the number of processors and other hardware resources would lead to a p-fold increase in the value of Sp. However, linear Sp is difficult to achieve, particularly as the number of processors is increased, due, e.g., to the fact that a larger number of processors usually implies a longer time spent with inter-processor communication.

The efficiency of a parallel system with p processors over a sequential system with a single processor can be defined as $Ef = Sp / p$. For instance, continuing the above example, where $Sp = 5$ and $p = 10$, the efficiency is $Ef = 5 / 10 = 0.5$.

11.3 Parallel EAs for Data Mining

One of the first steps in parallelizing a data mining algorithm (or any other kind of algorithm) should be to identify the "bottleneck" of the algorithm, i.e., which parts of the algorithm are most computationally expensive. In the case of EA this identification is simple. In general the bottleneck of an EA for data mining (particularly when mining very large data sets) is fitness evaluation, as mentioned above. In contrast, the application of genetic operators has a relatively cheap computational cost. Therefore, in this section we focus on how to parallelize fitness computation.

In a high level of abstraction, there are two basic approaches to exploit parallelism in fitness computation [Freitas and Lavington 1998, chap. 10]: a control-parallel approach or a data-parallel approach (subsection 11.2.3). These two approaches are discussed in the next two subsections (11.3.1 and 11.3.2), respectively. Subsection 11.3.3 discusses a hybrid control/data-parallel approach for parallelizing fitness computation.

To guide our discussion, in this chapter we will use the parallel EA taxonomy shown in Figure 11.4. The subtree rooted at the control-parallel approach is similar to parallel EA taxonomies often used in the literature – see, e.g., [Cantu-Paz 2000, chap. 1; Lee et al. 2000], since most parallel EAs follow a control-parallel approach. Figure 11.4 just extends this conventional taxonomy by adding on the top of it, at a higher-level of abstraction, the distinction between control-parallel, data-parallel and hybrid control/data-parallel approaches. We believe that, although this higher-level distinction is ignored in conventional taxonomies, it is very important in the context of data mining and other data intensive applications. Actually, exploiting data parallelism seems natural and necessary when mining very large data sets.

As shown in Figure 11.4, control-parallel EAs can be subdivided into single-population EAs and distributed-population EAs. The basic idea of the former is that selection and genetic operators are applied by considering a single population of individuals maintained by the EA. This approach is often implemented by a master-slave architecture, where a single master processor is in charge of performing selection and genetic operators and multiple slave processors are in charge of computing individuals' fitness in parallel. That is, the slaves receive individuals to be evaluated from the master, compute their fitness values in parallel, and return those values to the master, which then performs selection and genetic operators in a sequential fashion.

Note that, since the EA maintains a single population of individuals, the evolution of individuals in this kind of parallel EA will be the same as the evolution of individuals in its sequential counterpart, assuming of course that the same parameters are used in both versions of the EA. In particular, this kind of EA returns the same result as its sequential counterpart (running on a single-processor computer).

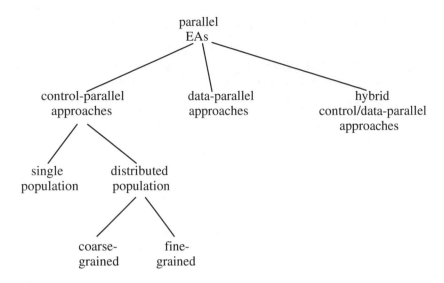

Figure 11.4: Taxonomy of parallel EAs

In distributed-population EAs the algorithm does not recognize a single population of individuals. Rather, individuals are distributed among multiple processors, and selection and genetic operators are essentially applied to subsets of individuals. As a result, it should be noted that the evolution of individuals in this kind of parallel EA will be different from the evolution of individuals in its sequential counterpart. Hence, unlike single-population parallel EAs, distributed-population parallel EAs return a result different from their sequential counterpart. This behavior of distributed-population EAs is not necessarily a bad thing. Actually, this behavior is sometimes used to reduce the chance that the EA will have a premature converge to a suboptimal solution [Giordana and Neri 1995].

Distributed-population EAs can be further subdivided into coarse-grained and fine-grained parallel EAs. In coarse-grained EAs the population is usually divided into a small number of subpopulations, each of them containing a large number of individuals. Each subpopulation is usually associated with a single processor. Selection and genetic operators are applied independently at each subpopulation, but the subpopulations occasionally communicate with each other in order to exchange individuals. Coarse-grained EAs are sometimes called multiple-population EAs or "island model" EAs.

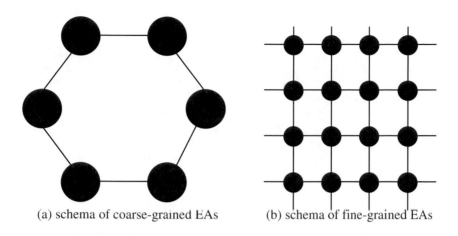

(a) schema of coarse-grained EAs   (b) schema of fine-grained EAs

Figure 11.5: Difference between coarse-grained and fine-grained parallel EAs

In fine-grained distributed EAs the population is divided into a large number of overlapping "subpopulations", each of them containing a small number of individuals. Actually, this kind of EA is often described as consisting of a single spatially-structured population, which is often structured in the form of a two-dimensional grid, as shown in Figure 11.5(b). (Figure 11.5(a) shows a simple schema of coarse-grained EAs, for the sake of comparison.) Selection and genetic operators are applied in parallel to small neighborhoods around each individual. The neighborhoods overlap, so that the genetic material of good individuals can gradually spread (or diffuse) across the entire spatial structure. We prefer to consider the overlapping neighborhoods as overlapping subpopulations, since selection and genetic operators are independently applied to each of those subpopulations. That is, the individual(s) assigned to a given processor and the individuals assigned to neighboring processors effectively constitute a "subpopulation", often called a "deme". In one extreme, ideally each processor will be assigned a single individual, so that all individuals have their fitness computed in parallel, at the same time. Fine-grained parallel EAs are sometimes called "diffusion model" EAs or "cellular" EAs (by analogy with cellular automata).

Intuitively, coarse-grained parallel EAs can be more naturally implemented in parallel computers with a relatively small number of processors (say on the order of tens of processors); whereas fine-grained parallel EAs can be more naturally implemented in massively-parallel computers with a large number of processors (say on the order of thousands of processors).

A more detailed discussion about control-parallel EAs in general, independent of data mining applications, can be found in a recent book by [Cantu-Paz 2000].

11.3.1 Exploiting Control Parallelism

One approach for parallelizing fitness computation consists of exploiting inter-individual (or inter-fitness-computation) parallelism. This approach is illustrated in Figure 11.6. In this approach the set of individuals of the EA is distributed across all the processors. Each processor computes the fitness of a different sub-set of individuals, and all processors work in parallel. This is a control-parallel approach (subsection 11.2.3), in the sense that the flow of control of the algorithm is parallelized. After all, at a given instant there are p individuals being evaluated in parallel, at the same time, by p different processors. However, this is a data-sequential approach. The computation of the fitness of each individual is done by a single processor, so that each individual's fitness is computed by accessing the data being mined in a sequential fashion, one instance at a time.

In the context of data mining and other data intensive applications, a purely control-parallel approach has some drawbacks concerning scalability to very large data sets. The basic problem is that in general the computation of the fitness of each individual requires access to the entire data being mined. Where should we store all this data? Two basic possibilities are as follows.

First, one can use a shared-memory system, so that all processors can compute the fitness of their corresponding individuals by directly accessing all data being mined, stored in the global, shared memory of the system. However, as mentioned above, shared-memory systems are not very scalable to a large number of processors.

Second, one can use a distributed-memory system, which is more scalable to a large number of processors. However, if the system has a large number of processors each processor would typically have a relatively small amount of local memory, so that the data being mined would have to be distributed across all processors' local memories (assuming the data being mined is very large). This would create the problem that there would be a high data traffic across the inter-process network during fitness computation, since each processor would need to access data stored in many other processors. (Of course, if each processor had a very large amount of local memory one could simply replicate the very large data set being mined in each processor' local memory, avoiding the just-mentioned high data traffic problem. However, in general the larger the amount of local memory per processor the smaller the number of processors of a parallel system.)

An example of a parallel GA for data mining exploiting control parallelism is GA-MINER [Flockhart and Radcliffe 1995]. This is a fine-grained parallel GA. GA-Miner was implemented on both a shared-memory system and a distributed-memory system. In the latter the entire data being mined was replicated in each processor's local memory, which reduces scalability for very large data sets, as mentioned above. (Indeed, in the experiments reported by the authors the distributed-memory system was used to mine a relatively small data set – with 2,557 data instances and 58 attributes.)

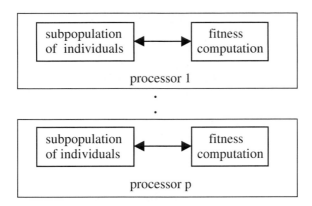

Figure 11.6: Exploiting control parallelism in fitness computation

Two other examples of parallel GAs for data mining exploiting control parallelism are REGAL [Giordana and Neri 1995; Neri and Giordana 1995; Neri and Saitta 1996] and its descendant G-NET [Anglano et al. 1997, 1998]. REGAL is a coarse-grained parallel GA. In essence, it consists of several Nodal GAs and a supervisor node. Each Nodal GA is assigned a subpopulation of individuals and a subset of training data instances, so that all Nodal GAs can work in parallel on different processors. Different Nodal GAs evolve different individuals (candidate classification rules) for classifying different instance subsets. This scheme realizes a kind of niching, where the rules evolved by different Nodal GAs cover different instance subsets, corresponding to different niches in data space.

The system also has a supervisor node that periodically collects the best local candidate rule found by each Nodal GA. (Note that each local candidate rule was evolved to cover the subset of data instances that was assigned to its corresponding Nodal GA.) Then the supervisor re-evaluates the current set of best rules on the entire data being mined. Note that this reduces scalability for very large data sets, as mentioned above. The result of this re-evaluation is used by the supervisor to dynamically re-assign data-instance subsets to Nodal GAs, in order to improve the current set of best candidate rules.

G-NET, a descendant of REGAL, is a fine-grained parallel GA. Hence, it is intuitively more suitable for parallel systems with a large number of processors than REGAL, which is a coarse-grained parallel GA. However, apparently G-NET still uses a supervisor node that periodically collects the best local candidate rule found by each genetic node and re-evaluates the current set of best rules on the entire data being mined (reducing scalability), in order to re-assign data instances to genetic nodes.

Another fine-grained parallel GA for data mining is proposed by [Llora and Garrell 2001a, 2001b]. This GA works with a population of individuals spread over a two-dimensional grid, so that each cell of the grid contains zero or one individual. However, in this work only a sequential implementation is reported.

11.3.2 Exploiting Data Parallelism

Intuitively, in the context of data mining involving very large data sets and other data-intensive applications, a promising approach to exploit parallelism in fitness computation consists of exploiting intra-individual (or intra-fitness-computation) parallelism. This approach is illustrated in Figure 11.7. In this approach the data being mined is distributed across the p processors of the parallel system. The computation of the fitness of each individual is performed in a data-parallel fashion, since at a given instant an individual's fitness is being evaluated in parallel, at the same time, by p different processors. However, this is a control-sequential approach, in the sense that the flow of control of the algorithm is sequential. After all, at a given instant there is just one individual having its fitness evaluated and individuals are processed on a one-individual-at-a-time basis, i.e., the computation of the fitness of an individual starts only after the computation of the fitness of the previous individual was completed.

In the context of mining very large data sets, the most important advantage of this kind of data-parallel approach is that, intuitively, it is considerably more scalable with respect to the size of the data being mined than a typical control-parallel approach. To put it in simple terms, the larger the size of the data being mined the higher the amount of parallelism to be exploited.

In addition, when using a distributed-memory system, the data-parallel approach avoids one major scalability problem associated with the control-parallel approach. Namely, in the data parallel approach the entire data being mined can be distributed among the local memories of all processors without causing a high data traffic in the inter-processor network during fitness computation, unlike the use of a control-parallel approach (discussed above).

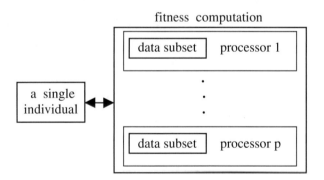

Figure 11.7: Exploiting data parallelism in fitness computation

But how does a data parallel approach avoid high data traffic on a distributed-memory system during the computation of the fitness of an individual? The answer is based on the fact that fitness evaluation can be performed in two steps, as follows:

(a) Each processor computes a partial fitness evaluation by accessing only the subset of data stored in its local memory;

(b) then the partial fitness evaluations performed by all the processors (which are very summarized, aggregated information, by comparison with the original data) are exchanged among all processors, in order to compute a global fitness evaluation.

To make our discussion somewhat more concrete, let us consider the case where each EA individual corresponds to a single prediction rule. In order to evaluate the fitness of an individual, assume one has to compute a confusion matrix for its corresponding rule, as explained in section 6.5. Recall that in the data-parallel approach being considered here the entire data being mined is distributed across the local memories of p processors of a distributed-memory system.

Hence, in order to compute the confusion matrix associated with an individual, at first each processor computes the partial values of TP, FP, FN, and TN (see section 6.5) by matching the individual's rule with the processor's local data set. Once all processors have completed this step, the total values of TP, FP, FN, and TN are computed by simply summing all the partial values for each of those four variables.

Figure 11.8 shows a very simple example involving only 2 processors, denoted p_1 and p_2, but this example can be straightforwardly generalized to a much larger number of processors. Figures 11.8(a) and 11.8(b) show the local, partial confusion matrices computed by processors p_1 and p_2 respectively, by accessing only their corresponding local data subsets. Figure 11.8(c) shows the global confusion matrix computed for the rule being evaluated. Each cell of this global matrix contains a value which is simply the sum of the two values in the corresponding cells of the two local confusion matrices.

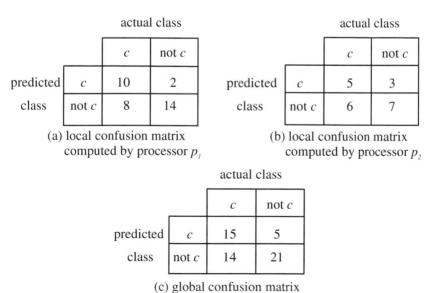

Figure 11.8: Example of distributed fitness evaluation in a data-parallel approach

Once a global confusion matrix has been computed, any of the fitness functions discussed in section 6.5 can be computed and then assigned to the individual being evaluated.

In the above discussion we have implicitly assumed that the application programmer is in charge of exploiting data parallelism in fitness evaluation. There is, however, an alternative way of exploiting data parallelism, which may be called automatic exploitation of data parallelism. In this case data parallelism issues are automatically handled by a parallel database server. The application program just sends database queries (say SQL queries) to the parallel database server. The query is automatically parallelized by the server.

In the context of data mining, the basic idea is to formulate database queries whose results contain all the necessary information to compute the confusion matrix of EA individuals. This approach is discussed in [Freitas 1997], in the context of GP, and in [Freitas and Lavington 1998, chap. 11], in the context of rule induction algorithms. (Although the context of the latter is rule induction rather than EAs, the basic ideas of query formulation are very similar in both paradigms.)

It should be noted that this approach introduces a client/server communication overhead (as well as a query-compilation/interpretation overhead) into the system. That is, the client application will have to send database queries to a parallel database server. In a networked database environment, where the queries have to be transmitted from the client to the server via a network, this overhead can be significant. This approach is recommended only when the amount of data being mined is very large, so that the client/server communication overhead is not significant by comparison with the time savings associated with parallel processing of database queries.

This approach has the advantage that it keeps all the data being mined in the parallel database server during the entire run of the EA. Only summarized results – necessary for computing confusion matrices of candidate rules (individuals) – are transmitted from the server to the client.

11.3.3 Exploiting Hybrid Control/Data-Parallelism

As mentioned above, both control parallelism and data parallelism can be combined in an EA for data mining. As one example, a GA for rule discovery exploiting both control and data parallelism was proposed by [Araujo et al. 1999, 2000]. In this GA each individual represents a single IF-THEN prediction rule, so that the GA follows the Michigan approach (subsection 6.1.1). The algorithm was designed to tackle the dependency modeling task. As discussed in section 2.2, this task can be regarded as an extension of the classification task where there are two or more goal attributes to be predicted. Hence, different individuals can represent rules with different goal attributes in their consequent (THEN part).

Hence, in this GA the consequent of a rule has the form "$G_i = V_{ij}$", where G_i is the i-th goal attribute and V_{ij} is the j-th value belonging to the domain of G_i. The population is divided into s subpopulations, each of them with N individuals. Selection and genetic operators such as crossover are independently applied within each subpopulation. In each subpopulation all N individuals represent rules with the same consequent (the same G_i and the same V_{ij}) during all the evolution of the GA. As discussed in subsection 6.1.3.4, once each subpopulation evolves independently from the others, this approach of associating a fixed rule consequent with each subpopulation avoids the exchange of genetic material between individuals (rules) being evolved to predict different goal attributes.

Each population is assigned to a distinct *logical* processor. Let p be the number of (*physical*) processors available in the parallel system. (The system could be a parallel machine or a network of workstations. The latter kind of parallel system was used by the authors.) We first discuss the case where $s = p$, i.e., the number of subpopulations equals the number of processors. The cases where $s < p$ or $s > p$ will be discussed later. We assume the use of a distributed-memory parallel system, where the parallelization of the algorithm is considerably more challenging, by comparison with a shared-memory parallel system (subsection 11.2.1).

When $s = p$ each subpopulation is allocated to a distinct processor, so that all processors are used. All subpopulations evolve in parallel. Not only the individuals being evolved by the GA, but also the data being mined is partitioned across all p processors. This data partitioning allows the exploitation of data parallelism, and it avoids the need for replicating the data being mined in all p processors. As mentioned above, such replication would considerably reduce scalability for large data sets in a distributed-memory system, where each processor usually has only a relatively small amount of local memory. The physical network of interconnection among the processors is mapped onto a logical ring of processors, so that each processor has both a right neighbor and a left neighbor.

At each generation, the fitness of all $s \times N$ individuals (where s is the number of subpopulations and N is the number of individuals per subpopulation) is computed by exploiting both control and data parallelism, as follows.

At first each processor computes a local confusion matrix (section 6.5) for each of the N individuals (rules) in its local subpopulation. This local confusion matrix is computed by accessing only the data subset stored in the processor's local memory (subsection 11.3.2). Then the individuals pass through all the processors in a kind of round-robin scheme, using the above-mentioned ring of processors. In other words, at each step of this round-robin scheme each processor transfer its entire local subpopulation of N individuals, as well as their local confusion matrix (or a partial fitness measure computed from that local confusion matrix), to its right neighbor. When a processor receives a subpopulation of N individuals from its left neighbor it performs the following tasks for each of the incoming individuals:

(a) It computes a local confusion matrix for the rule represented by that individual, accessing only its local data set;
(b) It combines this just-computed local confusion matrix (or a partial fitness measure computed from it) with the local confusion matrix (or partial fitness measure) previously associated with that individual, received from the left processor (a combination of two local confusion matrices was illustrated in Figure 11.8);

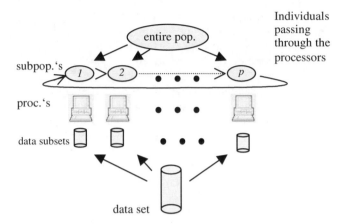

Figure 11.9: Round-robin scheme used to pass individuals through all processors

Then the processor forwards all the incoming individuals, with their corresponding updated confusion matrices (or partial fitness measure), to its right neighbor. This process, illustrated in Figure 11.9, is repeated until all individuals have passed through all processors and returned to their original processors. At this point the global confusion matrix (or final fitness measure) of all individuals has been computed. Note that at each step of the round-robin scheme only individuals and confusion matrices (or partial fitness measures), rather than data subsets, are exchanged among processors. This reduces the amount of interprocessor communication, which is desirable in parallel processing.

This approach exploits control parallelism because, at each step of the round-robin scheme, different processors are computing the fitness of different individuals. It also exploits data parallelism because the entire data being mined is partitioned across all processors and, within an entire cycle of the round-robin scheme (where each individuals passes through all processors), an individual has its fitness evaluated by all the p processors in a pipelined fashion (temporal parallelism). So, it can be considered a hybrid control/data-parallel approach.

Finally, once fitness evaluation is completed, each processor can apply selection and genetic operators to its local subpopulation of individuals independently from other processors. This phase exploits control parallelism.

The above discussion has assumed that $s = p$, i.e., the number of subpopulations equals the number of processors. If $s < p$ or $s > p$ the basic idea of the above arguments still holds, but some opportunity to exploit parallelism would probably be wasted. Firstly, suppose that $s < p$. In this case each subpopulation will be allocated to a distinct processor node, but some processors would be idle, reducing the efficiency in the exploitation of parallelism. Secondly, suppose that $s > p$. In this case the s subpopulations would be allocated to the p processors in a round-robin scheme, in order to ensure that the subpopulations will be as evenly distributed as possible across the available processors, achieving a good workload balance. In this case a physical processor would be in charge of evolving more than one subpopulation, underexploiting the potential for parallelism offered by the data mining task.

References

[Almasi and Gottlieb 1994] G.S. Almasi and A. Gottlieb. *Highly Parallel Computing*, 2nd edn. The Benjamim Cummings, 1994.

[Anglano et al. 1997] C. Anglano, A. Giordana, G. Lo Bello and L. Saitta. A network genetic algorithm for concept learning. *Proceedings of the 7th International Conference Genetic Algorithms*, 434–441. 1997.

[Anglano et al. 1998] C. Anglano, A. Giordana, G. Lo Bello and L. Saitta. An experimental evaluation of coevolutive concept learning. *Machine Learning: Proceedings of the 15th International Conference (ICML '98)*, 19–27. Morgan Kaufmann, 1998.

[Araujo et al. 1999] D.L.A. Araujo, H.S. Lopes and A.A. Freitas. A parallel genetic algorithm for rule discovery in large databases. *Proceedings of the 1999 IEEE Systems, Man and Cybernetics Conference, v. III, 940–945*. Tokyo, 1999.

[Araujo et al. 2000] D.L.A. Araujo, H.S. Lopes and A.A. Freitas. Rule discovery with a parallel genetic algorithm. In: A. Wu (Ed.) *Proceedings of the 2000 Genetic and Evolutionary Computation Conference (GECCO '2000) Workshop Program – Workshop on Data Mining with Evolutionary Algorithms*, 89–92. Las Vegas, NV, USA. 2000.

[Bhattacharrya 1998] S. Bhattacharrya. Direct marketing response models using genetic algorithms. *Proceedings of the 4th International Conference on Knowledge Discovery and Data Mining (KDD '98)*, 144–148. AAAI Press, 1998.

[Cantu-Paz 2000] E. Cantu-Paz. *Efficient and Accurate Parallel Genetic Algorithms*. Kluwer, 2000.

[Cavaretta and Chellapilla 1999] M.J. Cavaretta and K. Challapilla. Data mining using genetic programming: the implications of parsimony on generalization error. *Proceedings of the Congress on Evolutionary Computation (CEC '99)*, 1330–1337. IEEE, 1999.

[Flockhart and Radcliffe 1995] I.W. Flockhart and N.J. Radcliffe. GA-MINER: parallel data mining with hierarchical genetic algorithms - final report. *EPCC-AIKMS-GA-MINER-Report 1.0*. University of Edinburgh, UK, 1995.

[Freitas 1997] A.A. Freitas. A genetic programming framework for two data mining tasks: classification and generalized rule induction. *Genetic Programming 1997: Proceedings of the 2nd Annual Conference (GP '97)*, 96–101. Morgan Kaufmann, 1997.

[Freitas 1998] A.A. Freitas. A survey of parallel data mining. *Proceedings of the 2nd International Conference on the Practical Applications of Knowledge Discovery and Data Mining (PADD '98)*, 287–300. The Practical Application Company, London, 1998.

[Freitas and Lavington 1998] A.A. Freitas and S.H. Lavington. *Mining Very Large Databases with Parallel Processing*. Kluwer, 1998.

[Gathercole and Ross 1994] C. Gathercole and P. Ross. Dynamic training subset selection for supervised learning in genetic programming. *Parallel Problem Solving from Nature (PPSN-III)*, 312–321. Springer, 1994.

[Gathercole and Ross 1997a] C. Gathercole and P. Ross. Tackling the boolean even N parity problem with genetic programming and limited-error fitness. *Genetic Programming 1997: Proceedings of the 2nd Annual Conference (GP '97)*, 119–127. Morgan Kaufmann, 1997.

[Gathercole and Ross 1997b] C. Gathercole and P. Ross. Small populations over many generations can beat large populations over few generations in genetic programming. *Genetic Programming 1997: Proceedings of the 2nd Annual Conference (GP '97)*, 111–118. Morgan Kaufmann, 1997.

[Giordana and Neri 1995] A. Giordana and F. Neri. Search-intensive concept induction. *Evolutionary Computation 3(4)*: 375–416, 1995.

[Goldberg 1989] D.E. Goldberg. *Genetic Algorithms in Search, Optimization and Machine Learning*. Addison-Wesley, 1989.

[Hillis and Steele 1986] W.D. Hillis and L. Steele Jr. Data parallel algorithms. *Communications of the ACM, 29(12)*, 1170–1183, 1986.

[Lee et al. 2000] C.-H. Lee, S.-H. Park and J.-H. Kim. Topology and migration policy of fine-grained parallel evolutionary algorithms for numerical optimization. *Proceedings of the 2000 Congress on Evolutionary Computation (CEC '2000)*. IEEE, 2000.

[Lewis 1991] T.G. Lewis. Data parallel computing: an alternative for the 1990s. *IEEE Computer, 24(9)*, 110–111, 1991.

[Llora and Garrel 2001a] X. Llora and J.M. Garrell. Knowledge-independent data mining with fine-grained parallel evolutionary algorithms. *Proceedings of the Genetic and Evolutionary Computation Conference (GECCO '2001)*, 461–468. Morgan Kaufmann, 2001.

[Llora and Garrel 2001b] Inducing partially-defined instances with evolutionary algorithms. *Proceedings of the 18th International Conference on Machine Learning (ICML '2001)*, 337–344. Morgan Kaufmann, 2001.

[Neri and Giordana 1995] F. Neri and A. Giordana. A parallel genetic algorithm for concept learning. *Proceedings of the 6th International Conference Genetic Algorithms*, 436–443. 1995.

[Neri and Saitta 1996] F. Neri and L. Saitta. Exploring the power of genetic search in learning symbolic classifiers. *IEEE Transactions on Pattern Analysis and Machine Intelligence*, 18(11), 1135–1141, 1996.

[Papagelis and Kalles 2001] A. Papagelis and D. Kalles. Breeding decision trees using evolutionary techniques. *Proceedings of the 18th International Conference on Machine Learning (ICML '2001)*, 393–400. Morgan Kaufmann, 2001.

[Sharpe and Glover 1999] P.K. Sharpe and R.P. Glover. Efficient GA based techniques for classification. *Applied Intelligence 11*, 277–284, 1999.

[Teller and Andre 1997] A. Teller and D. Andre. Automatically choosing the number of fitness cases: the rational allocation of trials. *Genetic Programming 1997: Proceedings of the 2nd Annual Conference (GP '97)*, 321–328. Morgan Kaufmann, 1997.

12 Conclusions and Research Directions

> "Each new discovery in science has the role of a
> small piece of cloth, which belongs to a beautiful,
> colourful, big, unfinished patchwork quilt. We,
> human beings, uselessly try to finish it, as a
> way of understanding our own existence."
> Pierre Simon (Marquis of Laplace), 1749-1827

This chapter is divided into two parts. Section 12.1 presents some general remarks on data mining with evolutionary algorithms (EAs). These remarks can be regarded as a very compact summary of the main arguments of this book. They discuss, in general, the suitability of EAs for data mining with respect to the issues of predictive accuracy and comprehensibility of discovered knowledge, as well as the issue of the computational time taken by EAs.

Section 12.2 proposes three research directions, namely developing EAs for data preparation, developing multi-objective EAs for data mining, and developing a genetic programming system for algorithm induction.

12.1 General Remarks on Data Mining with EAs

12.1.1 On Predictive Accuracy

With respect to the predictive accuracy of discovered knowledge, one can argue that in general EAs are roughly competitive (sometimes somewhat better, sometimes somewhat worse) than conventional rule induction and decision tree induction algorithms.

It should be noted that this argument tends to be true for any algorithm of any other data mining or machine learning paradigm, and not only for EAs. The point that the predictive accuracy of a data mining or machine learning algorithm is strongly dependent on the application domain (more precisely, the data set being mined) is now well-established. It has been shown both theoretically [Schaffer 1994; Rao et al. 1995; Schaffer 1993] and empirically [Michie et al. 1994; King et al. 1995].

What is important is to understand in which cases – for which kinds of data set – the predictive accuracy of an EA will tend to be better or worse than the predictive accuracy of a conventional rule induction or decision tree induction algorithm. Although a precise answer for this question is still unknown, as a first approximation one can say that, intuitively, the larger the degree of attribute interaction in the data being mined, the more an EA's predictive accuracy tends to outperform the predictive accuracy of a greedy, local-search-based rule induction or decision tree induction algorithm – see subsection 1.3.2 and in particular subsection 3.2.1.

12.1.2 On Knowledge Comprehensibility

In order to discuss the comprehensibility (or interpretability) of discovered knowledge, we should recall that EAs are a very general search paradigm, and they can be used for data mining and knowledge discovery in many different ways, varying from a method to optimize some parameters of another data mining algorithm – e.g., to optimize the attribute weights of a nearest neighbor algorithm – to a method used to directly discover prediction rules or another form of knowledge.

In the former case, of course, the comprehensibility of knowledge discovered by an EA is not a relevant issue, since the EA is just being used as an auxiliary parameter-optimization method, rather than as a knowledge discovery method.

In the latter case, the comprehensibility of the knowledge discovered by an EA is, of course, strongly dependent on the knowledge representation associated with an individual. As long as an individual represents a rule, rule set or decision tree, the EA can discover knowledge as comprehensible as the knowledge discovered by a conventional rule induction or decision tree induction algorithm.

Actually, a large part of this book was devoted to show this point. Both chapters 6 and 7 as a whole, as well as some sections of chapter 10, have focused on the use of EAs for rule discovery.

Furthermore, EAs for rule discovery can directly use rule pruning methods (which make a rule simpler by removing some conditions from it) similar to the ones often used by conventional rule induction algorithms – see subsection 6.2.3. After all, the issue of applying a rule pruning operator to a given rule is orthogonal to the issue of how that rule was generated (by evolutionary or non-evolutionary algorithms).

12.1.3 On Computational Time

It can be argued that, in general, EAs are one or two orders of magnitude slower than conventional rule induction and decision tree algorithms. This is a disadvantage of EAs. However, it should be noted that there are some ways of mitigating this problem, as discussed in chapter 11.

That chapter discussed two general approaches for scaling up EAs for large data sets. The first one consists of using only data subsets (rather than the entire training set) in fitness evaluation. The second approach consists of using parallel processing techniques to speed up fitness evaluation. The first approach is usually easier to implement, particularly for programmers not familiar with parallel processing. However, it might lead to some degradation in predictive accuracy in some cases [Freitas and Lavington 1998, pp. 92-97].

The parallel processing approach, on the other hand, can be used to speed up an EA without sacrificing predictive accuracy, since one can develop a parallel EA that discovers the same prediction rules as its sequential counterpart – assuming, of course, that both the parallel and sequential programs produce the same sequence of randomly-generated numbers to implement stochastic genetic operators. In particular, very large data sets seem to offer both a need and an opportunity for exploiting data parallelism or hybrid data/control parallelism, as discussed in subsections 11.3.2 and 11.3.3.

In any case, it should be noted that in some applications the need for speeding up EAs is not as important as it seems at first glance. Data mining is typically an off-line task, and in general the time spent with running a data mining algorithm is a small fraction (less than 20%) of the total time spent with the entire knowledge discovery process (see section 1.1). Data preparation is usually the most time consuming phase of this process. Hence, in many applications, even if a data mining algorithm is run for several hours or several days, this can be considered an acceptable processing time, at least in the sense that it is not the bottleneck of the knowledge discovery process.

12.2 Research Directions

12.2.1 Developing EAs for Data Preparation

In chapter 9 we have discussed several EAs for data preparation – including mainly attribute selection, attribute weighting, and attribute construction (or constructive induction).

We believe that, in the general area of data preparation for data mining, the most promising area for future research is attribute construction. As mentioned in section 4.3, the motivation for attribute construction is particularly strong when the original attributes do not have much predictive power by themselves. In this case, the construction of new attributes with greater predictive power than the original attributes tends to be even more important than the classification algorithm itself. In other words, it is often the case that improving the data representation is more important than improving the classification algorithm.

Intuitively, a significant challenge in the construction of new attributes is how to effectively cope with attribute interactions [Freitas 2001]. As mentioned above, an advantage of EAs for data mining is precisely that they tend to cope better with attribute interaction, by comparison with most local greedy search algorithms that are often used in rule induction and attribute construction.

Although there has already been extensive research on EAs for attribute selection (see section 9.1), we believe there is still some opportunity and some need for further research in this area, as follows. First of all, it should be noted that the vast majority of EAs for attribute selection follow the wrapper approach. As discussed in subsection 4.1.2.1, in general the wrapper approach is much more computationally expensive than the filter approach, so that the former tends to be impractical for large data sets. Hence, one needs more research on EAs for attribute selection following the filter approach. Some first steps towards this direction can be found in [Chen and Liu 1999] (subsection 9.1.2) and [Martin-Bautista and Vila 1998].

12.2.2 Developing Multi-objective EAs for Data Mining

In the context of rule discovery, there are different objectives to be optimized by an EA, such as a rule's predictive accuracy, simplicity (or comprehensibility) and interestingness – see subsection 1.1.1 and section 2.3.

We have briefly argued at the end of section 6.5 that these objectives involve incommensurable rule quality criteria, and that they can represent conflicting goals – at least in some application domains.

This suggests the development of multi-objective EAs for rule discovery, where the fitness of each individual (a rule or rule set) is not a single scalar value, but rather a vector of two or more values, where each value measures a different aspect of rule quality. One of the values should probably be a measure of predictive accuracy, while the other value(s) could be a measure of simplicity or/and interestingness. Hence, the EA would search for non-dominated solutions (in the Paretto sense) with respect to the values of the fitness vector [Deb 2001].

Some recently proposed multi-objective EAs for data mining and knowledge discovery are described in [Bhattacharyya 2000a, 2000b], which focus on a marketing application, and in [Kim et al. 2000; Emmanouilidis et al. 2000], which involve EAs for attribute selection.

It should be noted that EAs seem to be well-suited for multi-objective tasks in general, since they already work with a population of individuals (candidate solutions) – and in multi-objective search we do have to find a set of solutions, rather than a single solution. This can be considered a good indication that developing multi-objective EAs for data mining is a promising direction for future research.

12.2.3 Developing a GP System for Algorithm Induction

We have argued, at the end of chapter 7, that the power of genetic programming (GP) for data mining has been under-explored in the literature, as follows. In that chapter we have discussed several GP systems that induce a high-level classifier, expressed in the form of prediction rules or decision trees, from data. However, in every case, the GP was designed and used for inducing a classifier for one data set at a time. In other words, the result of a GP run was a classifier for a given data set, in a particular application domain. That classifier could be called a "program" only in a loose sense of the word.

Intuitively, GP is about the automatic discovery of computer programs, or algorithms, that are general "recipes" for solving a given kind of problem regardless of the application domain. For instance, classification is one kind of problem, regardless of the application domain – medical applications, financial applications, manufacturing applications, etc. Most classification algorithms are generic enough to be applied to different application domains, but virtually all current classification algorithms – including GP systems – were designed by a human being. The general "recipe" was written by a human. What virtually any current classification algorithm does is to apply a human-written recipe to a particular data set. Perhaps one can say that the credit for the great flexibility of current classification algorithms, which allows them to be applied to many different application domains, should be assigned mainly to the human designers of those algorithms, rather than to the artificial "intelligence" of the computer.

We believe that GP has a good potential to significantly increase the degree of automation – and so increase the degree of artificial intelligence – in data mining. It seems possible, at least in principle, to use GP to actually discover a generic data mining (say, a classification) algorithm. This approach may be called algorithm induction, rather than rule induction. We do not have any good solution for this problem yet, but we will suggest here some intriguing possibilities, which hopefully will inspire other researchers to pursue this challenging, ambitious research direction.

The basic idea is to recall that GP works not only with a terminal set – containing attributes and attribute values, in the context of data mining – but also with a function set – containing functions or operators that can be applied to the attributes or attribute values of the terminal set. Now, in principle there is no reason to limit the contents of the function set to the simple operators (AND, OR, "+", "-", "≤", ">", etc.) often used in most data mining algorithms.

The function set of GP can contain much higher-level operations, suitable for algorithm induction, rather than rule induction. The kind of operations that we have in mind are, for example, generic operations such as *generalize(r), specialize(r), compute-class-distribution(r)*, as well as some general program-control structures such as IF-THEN. In this example, *generalize(r)* and *specialize(r)* are operations that perform the generalization or specialization of a given candidate rule *r*, whereas *compute-class-distribution(r)* would compute the class distribution of the data instances covered by rule *r*. Hence, we can roughly envisage an individual of GP combining these operations in a logical sequence that forms a sensible rule induction algorithm, something of the form:

IF *compute-class-distribution(r)* = "a-certain-kind-of-result"
THEN *generalize(r)*
ELSE *specialize(r)* .

Clearly, using this kind of higher-level operations in the function set is just part of the solution. Several other "tricks" will have to be used. For instance, since the goal would be to evolve a general-purpose data mining algorithm, intuitively, in order to evaluate an individual, the training set would have to contain data from several different data sets, from different application domains.

Overall, this goal would be very ambitious, but we believe it is worthwhile its investigation, since, if successful, it would probably be a major advance in data mining research, as well as in artificial intelligence and artificial programming research.

References

[Bhattacharyya 2000a] S. Bhattacharyya. Evolutionary algorithms in data mining: multi-objective performance modeling for direct marketing. *Proceedings of the 6th ACM SIGKDD International Conference on Knowledge Discovery and Data Mining (KDD '2000)*, 465–473. ACM, 2000.

[Bhattacharyya 2000b] S. Bhattacharyya. Multi-objective data mining using genetic algorithms. In: A.S. Wu (Ed.) *Proceedings of the 2000 Genetic and Evolutionary Computation Conference Workshop Program – Worshop on Data Mining with Evolutionary Algorithms*, 76–79. 2000.

[Chen and Liu 1999] K. Chen and H. Liu. Towards an evolutionary algorithm: a comparison of two feature selection algorithms. *Proceedings of the Congress on Evolutionary Computation (CEC '99)*, 1309–1313. Washington D.C., USA, 1999.

[Deb 2001] K. Deb. *Multi-Objective Optimization Using Evolutionary Algorithms*. Wiley, 2001.

[Emmanouilidis et al. 2000] C. Emmanouilidis, A. Hunter and J. MacIntyre. A multiobjective evolutionary setting for feature selection and a commonality-based crossover operator. *Proceedings of the 2000 Congress on Evolutionary Computation (CEC '2000)*, 309–316. IEEE, 2000.

[Freitas 2001] A.A. Freitas. Understanding the Crucial Role of Attribute Interaction in Data Mining. *Artificial Intelligence Review 16(3)*, 177–199, 2001.

[Freitas and Lavington 1998] A.A. Freitas and S.H. Lavington. *Mining Very Large Databases with Parallel Processing*. Kluwer, 1998.

[Kim et al. 2000] Y. Kim, W.N. Street and F. Menczer. Feature selection in unsupervised learning via evolutionary search. *Proceedings of the 6th ACM SIGKDD International Conference on Knowledge Discovery and Data Mining (KDD '2000)*, 365–369. ACM, 2000.

[King et al. 1995] R.D. King, C. Feng and A. Sutherland. STATLOG: comparison of classification algorithms on large real-world problems. *Applied Artificial Intelligence*, 9(3), 289–333, 1995.

[Martin-Bautista and Vila 1998] M.J. Martin-Bautista and M.-A. Vila. Applying genetic algorithms to the feature selection problem in information retrieval. *Flexible Query Answering Systems (Prod. 3rd International Conference, FQAS' 98). Lecture Notes in Artificial Intelligence 1495*, 272–281. Springer, 1998.

[Michie et al. 1994] D. Michie, D.J. Spiegelhalter and C.C. Taylor. *Machine Learning, Neural and Statistical Classification*. Ellis Horwood, New York, 1994.

[Rao et al. 1995] R.B. Rao, D. Gordon and W. Spears. For every generalization action, is there really an equal and opposite reaction? Analysis of the conservation law for generalization performance. *Proceedings of the 12th International Conference Machine Learning (ML '95)*, 471–479. 1995.

[Schaffer 1993] C. Schaffer. Overfitting avoidance as bias. *Machine Learning* 10, 153–178, 1993.

[Schaffer 1994] C. Schaffer. A conservation law for generalization performance. *Proceedings of the 11th International Conference Machine Learning*, 259–265. 1994.

Index

Natural Computing Series

W.M. Spears: Evolutionary Algorithms. The Role of Mutation and Recombination. XIV, 222 pages, 55 figs., 23 tables. 2000

H.-G. Beyer: The Theory of Evolution Strategies. XIX, 380 pages, 52 figs., 9 tables. 2001

L. Kallel, B. Naudts, A. Rogers (Eds.): Theoretical Aspects of Evolutionary Computing. X, 497 pages. 2001

M. Hirvensalo: Quantum Computing. XI, 190 pages. 2001

G. Păun: Membrane Computing. An Introduction. XI, 429 pages, 37 figs., 5 tables. 2002

A.A. Freitas: Data Mining and Knowledge Discovery with Evolutionary Algorithms. XIV, 264 pages, 74 figs., 10 tables. 2002

H.-P. Schwefel, I. Wegener, K. Weinert (Eds.): Advances in Computational Intelligence. VIII, 325 pages. 2002

A. Ghosh, S. Tsutsui (Eds.): Advances in Evolutionary Computing. Approx. 980 pages. 2002

L.F. Landweber, E. Winfree (Eds.): Evolution as Computation. DIMACS Workshop, Princeton, January 1999. Approx. 300 pages. 2002

M. Amos: Theoretical and Experimental DNA Computation. Approx. 200 pages. 2002

Printed by Publishers' Graphics LLC